Gero Brockschnieder

Asymptotics of Cubic Number Fields with Bounded Second Successive Minimum of the Trace Form

Anchor Academic
Publishing

Brockschnieder, Gero: Asymptotics of Cubic Number Fields with Bounded Second
Successive Minimum of the Trace Form, Hamburg, Anchor Academic Publishing 2015

Buch-ISBN: 978-3-95489-389-8
PDF-eBook-ISBN: 978-3-95636-336-8
Druck/Herstellung: Anchor Academic Publishing, Hamburg, 2015

Bibliografische Information der Deutschen Nationalbibliothek:
Die Deutsche Nationalbibliothek verzeichnet diese Publikation in der Deutschen
Nationalbibliografie; detaillierte bibliografische Daten sind im Internet über
http://dnb.d-nb.de abrufbar.

Bibliographical Information of the German National Library:
The German National Library lists this publication in the German National Bibliography.
Detailed bibliographic data can be found at: http://dnb.d-nb.de

All rights reserved. This publication may not be reproduced, stored in a retrieval system
or transmitted, in any form or by any means, electronic, mechanical, photocopying,
recording or otherwise, without the prior permission of the publishers.

———————————

Das Werk einschließlich aller seiner Teile ist urheberrechtlich geschützt. Jede Verwertung
außerhalb der Grenzen des Urheberrechtsgesetzes ist ohne Zustimmung des Verlages
unzulässig und strafbar. Dies gilt insbesondere für Vervielfältigungen, Übersetzungen,
Mikroverfilmungen und die Einspeicherung und Bearbeitung in elektronischen Systemen.

Die Wiedergabe von Gebrauchsnamen, Handelsnamen, Warenbezeichnungen usw. in
diesem Werk berechtigt auch ohne besondere Kennzeichnung nicht zu der Annahme,
dass solche Namen im Sinne der Warenzeichen- und Markenschutz-Gesetzgebung als frei
zu betrachten wären und daher von jedermann benutzt werden dürften.

Die Informationen in diesem Werk wurden mit Sorgfalt erarbeitet. Dennoch können
Fehler nicht vollständig ausgeschlossen werden und die Diplomica Verlag GmbH, die
Autoren oder Übersetzer übernehmen keine juristische Verantwortung oder irgendeine
Haftung für evtl. verbliebene fehlerhafte Angaben und deren Folgen.

Alle Rechte vorbehalten

© Anchor Academic Publishing, Imprint der Diplomica Verlag GmbH
Hermannstal 119k, 22119 Hamburg
http://www.diplomica-verlag.de, Hamburg 2015
Printed in Germany

Contents

List of Figures   v

List of Tables   vi

List of Algorithms   vii

Abstract (English)   viii

Abstract (German)   x

1. Introduction   1

2. Preliminaries   6
   - 2.1. The Theory of Number Fields 6
   - 2.2. Class Field Theory 12
   - 2.3. Additional Preparations 14

3. A Finite Set of Polynomials   15
   - 3.1. Preparations 15
   - 3.2. Counting Polynomials 18
   - 3.3. Estimation of Errors 22

4. Parametrization of the Polynomials   26
   - 4.1. The Minimality of a Pair (B, C) 31
   - 4.2. A Bound for the Number of Galois Fields 39
   - 4.3. An Alternative Approach to the Bound for the Number of Galois Fields 45
   - 4.4. Fields with more than one related Minimal Pair (B, C) 47

5. The Rate of Convergence   **53**
    5.1. The Main Algorithm 54
    5.2. The Class Number Formula and Other Convergences 56
    5.3. Technical Details and an Overview of the Computational Results 59

6. Summary and Outlook   **62**

A. Computational Results   **66**

B. Source Codes   **68**

References   **71**

List of Figures

1. Value of the fraction $\frac{|P(X)| \cdot X^{-5/2}}{\sqrt{6/15}}$ 20
2. Value of the fraction $\frac{\rho X}{X^{5/2}}$ 57
3. Value of the fraction $10^2 \cdot \frac{\kappa X^{5/2} - |F(X)|}{X^2}$ 59
4. Running time of Algorithm 2 in milliseconds 61
5. Graphs of $|F(X)|$ and $\kappa X^{5/2}$ 63
6. Value of the fraction $\frac{\kappa X^{5/2}}{|F(X)|}$ 64

List of Tables

1. Value of the fraction $\frac{|P(X)| \cdot X^{-5/2}}{\sqrt{6/15}}$ 20
2. Value of the fraction $\frac{\rho X}{X^{5/2}}$ 58
3. Value of the fraction $\frac{\kappa X^{5/2} - |F(X)|}{X^2}$ 59
4. Overview: Computational results 60
5. Running time of Algorithm 2 61
6. Size of the set $P(X)$ for larger values of X 63
7. Computational results 67

List of Algorithms

1. Compute the number of fields K with $M_2(K) \leq X$ 54
2. Compute the number of fields K with $M_2(K) \leq X$ together with ρ_X 55

Abstract (English)

We present a new way of investigating totally real algebraic number fields of degree 3. Instead of making tables of number fields with restrictions only on the field discriminant and/or the signature as described by Pohst, Martinet, Diaz y Diaz, Cohen, and other authors, we bound not only the field discriminant and the signature but also the second successive minima of the trace form on the ring of integers $\mathcal{O}_K$ of totally real cubic fields K. With this, we eventually obtain an asymptotic behaviour of the size of the set of fields which fulfill the given requirements. This asymptotical behaviour is only subject to the bound X for the second successive minima, namely the set in question will turn out to be of the size $\mathcal{O}(X^{5/2})$.

We introduce the necessary notions and definitions from algebraic number theory, more precisely from the theory of number fields and from class field theory as well as some analytical concepts such as (Riemann and Dedekind) zeta functions which play a role in some of the computations. From the boundedness of the second successive minima of the trace form of fields we derive bounds for the coefficients of the polynomials which define those fields, hence obtaining a finite set of such polynomials. We work out an elaborate method of counting the polynomials in this set and we show that errors that arise with this procedure are not of important order. We parametrize the polynomials so that we have the possibility to apply further concepts, beginning with the notion of minimality of the parametrization of a polynomial. Considerations about the consequences of allowing only minimal pairs (B, C) (as parametrization of a polynomial $f(t) = t^3 + at^2 + bt + c$) to be of interest as well as a bound for the number of Galois fields among all fields in question and their importance in the procedure

Abstract (English) ix

of counting minimal pairs, polynomials, and fields finally lead to the proof that the number of fields K with second successive minimum $M_2(K) \leq X$ divided by the size of the suitably "cut back" set of polynomials tends to 1 if X tends to infinity, particularly because the number of fields with more than one related minimal pair (B, C) is of negligible order.

A considerable amount of work accounts for the computational investigation of the theory, namely we show how fast the convergence of the above-mentioned limit actually is by computing the value of the fraction for several values of X. Computational results are presented as comprehensive tables and, as a vivid representation, as graphs.

Abstract (German)

Es wird eine neue Art, total reelle algebraische Zahlkörper vom Grad 3 zu untersuchen, vorgestellt. Anstatt Tabellen von Zahlkörpern mit beschränkter Diskriminante und/oder Signatur zu erstellen (wie Pohst, Martinet, Diaz y Diaz, Cohen und andere Autoren), beschränken wir außerdem die zweiten sukzessiven Minima der Spurform auf $\mathcal{O}_K$, dem Ring der ganzen Zahlen total reeller kubischer Zahlkörper K. Es wird ein asymptotisches Verhalten für die Kardinalität der Menge aller Körper, die diese Voraussetzungen erfüllen, hergeleitet. Die Asymptotik ist dabei nur abhängig von der Schranke X für die zweiten sukzessiven Minima und es wird gezeigt, dass die Kardinalität der untersuchten Menge $\mathcal{O}(X^{5/2})$ ist.

Wir führen die notwendigen Begriffe und Definitionen aus der algebraischen Zahlentheorie, speziell aus der Theorie der Zahlkörper und der Klassenkörpertheorie, ein. Weiterhin werden einige analytische Konzepte wie (Riemannsche und Dedekindsche) Zetafunktionen, die in einigen Berechnungen eine Rolle spielen, eingeführt. Aus der Schranke für die zweiten sukzessiven Minima der Spurform der Körper werden Schranken für die Koeffizienten der Polynome, die diese Körper beschreiben, hergeleitet und als Resultat eine endliche Menge solcher Polynome beschrieben. Es wird beschrieben, wie die Elemente dieser Menge schnell und sinnvoll gezählt werden können, dabei wird gezeigt, dass die dabei auftretenden Fehler asymptotisch nicht von Bedeutung sind. Die Polynome werden parametrisiert, so dass weitere Konzepte angewandt werden können, allen voran wird die Minimalität einer Parametrisierung eines Polynoms eingeführt. Nachdem beschrieben wurde, warum nur minimale Paare (B, C) (als Parametrisierung eines Polynoms $f(t) = t^3 + at^2 + bt + c$) von

Abstract (German)

Interesse sind und eine Schranke für die Anzahl der Galoiserweiterungen unter den fraglichen Körpern sowie deren Bedeutung beim Zählen von minimalen Paaren, Polynomen und Körpern hergeleitet wurde, wird schließlich gezeigt, dass die Anzahl der Körper K mit zweitem sukzessiven Minimum $M_2(K) \leq X$ dividiert durch die Kardinalität der passend "beschnittenen" Menge der Polynome für $X \to \infty$ gegen 1 konvergiert, was unter anderem daraus folgt, dass die Anzahl der Körper mit mehr als einem zugehörigen minimalen Paar (B, C) asymptotisch keine Rolle spielt.

Ein nicht unerheblicher Aufwand kommt der rechnerischen Untersuchung der Theorie zugute. Es wird gezeigt, wie schnell die Konvergenz des genannten Grenzwertes ist, indem der Wert des Bruchs für verschiedene X berechnet wird. Die Ergebnisse werden als ausführliche Tabellen und, zur Veranschaulichung, als Graphen dargestellt.

1. Introduction

Algebraic number fields, particularly of small degree n, have been treated in detail in several publications during the last years. The two textbooks by Cohen (1993 and 2000), for example, offer lots of information about the topic itself as well as a comprehensive list of references. The subject that has been investigated the most is the computation of lists of number fields K with field discriminant $d(K)$ less than or equal to a given bound D and the computation of the minimal value of the discriminant for a given degree n (and often also signature (r_1, r_2)) of the number fields. The distinct cases of different degrees, as well as the different numbers of real and complex embeddings, respectively, are usually treated independently of each other since each case itself offers a broad set of problems and questions. Comprehensive lists and many values have been computed for the cases of fixed degree $n \leq 8$ in Buchmann and Ford, 1989 (for $n = 4$), Buchmann et al., 1993 (also for $n = 4$), Pohst, 1975 (for $n = 5, 6, 7, 8$), Pohst, 1982 (for $n = 6$) and Pohst et al., 1990 (for $n = 8$) where some of them concentrate on only some choices for the signature while others cover all of them. In some of the cases the applied methods and algorithms have been notably improved over the years.

Each value for the degree n of the investigated fields represents a huge and interesting set of problems and questions that can be treated on its own. The case we will concentrate on in this thesis is $n = 3$. Algebraic number fields of degree 3 are often referred to as *cubic fields* and, in a way, their investigation is easier than the investigation of higher degree fields since the higher the degree of the field, the higher the number of possible signatures (i.e. combinations of real and complex embeddings of the field). Usual questions regarding cubic fields are, as in other cases, the number of fields with discriminant less than or equal to a given bound as treated in Cohen, 2000, and the computation of the density of discriminants of those fields as treated in Roberts, 2000, or, much earlier, in Davenport and Heilbronn, 1969, and Davenport and Heilbronn, 1971.

In this thesis, we will concentrate only on *totally real cubic fields*. Totally real fields are those fields K for which each embedding of K into the complex numbers $\mathbb{C}$ has an image that lies inside the real numbers $\mathbb{R}$. An important result will be based on the work of Cohn, 1954, on the density of abelian (and hence totally real) cubic fields.

The purpose of this thesis is to show that the number of isomorphism classes of cubic fields K whose second successive minima $M_2(K)$, as introduced by Minkowski, are less than or equal to a given bound X is asymptotically equal (in X) to the number of cubic polynomials defining these fields modulo a relation $\sim_P$ which will be explained in detail. Loosely speaking, we will investigate how to "cut back" the set of polynomials, which is bounded only by the choice of X and a restriction on the coefficient of the quadratic term of the polynomials, to obtain a new set of polynomials which has asymptotically the same size as the set of isomorphism classes of cubic fields generated by polynomials of the original set. Usually one would expect a finite set of polynomials P whose coefficients are bounded by suitable values to be somewhat larger than the set F of all isomorphism classes of cubic number fields generated by these polynomials. The inequality $|F| \leq |P|$ is clear although a general statement about the difference of the two cardinalities cannot be made that simply.

The problems that have to be dealt with in the course of our investigation are: how to count the polynomials in the set defined by the bound X and the restriction on the coefficient of the quadratic term effectively? How to treat the errors that arise when counting the polynomials in a more elaborate way than just counting them one by one? How to represent the polynomials in a way so that a universal correspondence between polynomials and isomorphism classes of cubic fields can be introduced? Do we have to treat Galois fields different from fields that are non-Galois?

1. Introduction

The thesis will treat these questions in reasonable order by investigating three sets and mappings between them. The sets we will deal with are

$$F(X) = \{\text{Isomorphism classes of cubic fields } K \mid K \text{ is totally real}, M_2(K) \leq X\},$$
$$P(X) = \{f(t) = t^3 + at^2 + bt + c \mid a, b, c \in \mathbb{Z}, 0 \leq a \leq 2, 0 < a^2 - 2b \leq X, d(f) > 0, f \text{ is irreducible}\}, \text{ and}$$
$$P_\kappa(X) = \{(B, C) \mid (B, C) \text{ is a minimal parametrization of an irreducible polynomial } f \text{ with } B, C > 0\}$$

where the notion of a minimal pair (B, C) as parametrization of a polynomial f contains some inequalities which bound the set $P_\kappa(X)$. The two mappings we will investigate are

$$P(X) \to F(X) \text{ and}$$
$$P_\kappa(X) \to F(X).$$

The first result regarding these sets is a statement about the size of the set $P(X)$ which will turn out to be $\frac{\sqrt{6}}{15}X^{5/2} + \mathcal{O}(X^{3/2})$. The set $P_\kappa(X)$ is somewhat smaller than $P(X)$ and its size will turn out to be asymptotically equal to $\frac{1}{2}\zeta(5)^{-1}\frac{\sqrt{6}}{15}X^{5/2}$ with the factor $\frac{1}{2}$ being due to the facts that we always have $B > 0$ and that (B, C) and $(B, -C)$ always belong to fields of the same isomorphism class (which is also the reason we allow only pairs (B, C) with $B, C > 0$ in $P_\kappa(X)$). During the investigation of $P_\kappa(X)$ we will find out that the number of Galois fields among the set of all fields "belonging" to pairs $(B, C) \in P_\kappa(X)$ (which imply that the related pair (B, C) is counted three times instead of only once as it is the case for non-Galois fields) is negligible, namely it is $\mathcal{O}(X^{3/2})$. Furthermore we will bound the number of fields with more than one related pair (B, C) to $\mathcal{O}(X^2)$ so that they do not play a role in the asymptotical consideration as well. Putting all these results together, we will prove that, based on the set $P(X)$, we obtain the set $P_\kappa(X)$ which is asymptotically as "large" as the set $F(X)$. Attention should be paid to the fact that $|F(X)|$ cannot be determined easier than by counting all its elements, which means that there is no closed formula for this cardinality only subject to the value of X. However, via the cardinality of $P_\kappa(X)$, which we will get to know asymptotically, we can make a statement about $|F(X)|$, which is also asymptotic. It makes sense to investigate and illustrate this asymptotical behavior computationally, which will be done as a supplement and, in a way, as an application of the theory.

We will begin with some preliminaries to introduce the notions and concepts that we will run across repeatedly throughout the whole thesis in Chapter 2 where we give a short introduction into the most important notions in the theory of number fields. Some concepts from class field theory will later be important and we give the necessary definitions. In Chapter 3 we will build the set of polynomials we are interested in by evaluating the consequences of the few requirements we have on the polynomial coefficients which will eventually lead to a finite set of polynomials of size $\mathcal{O}(X^{5/2})$. We will see that we can count the number of polynomials in this set faster than just to count them one by one and we will evaluate the errors that arise when doing that. Furthermore, as a first important result, we will see that the number of reducible polynomials we have to think about is of negligible order. In order to obtain a correspondence between polynomials and fields, we introduce a parametrization of the polynomials in Chapter 4 where we will also see, as an important result, that the number of Galois fields in question is negligible which implies the desired correspondence. This will finally enable us to proof the assumption that the number of fields in question is asymptotically equal to the number of polynomials defining these fields modulo the relation $\sim_P$ which, in a way, "cuts back" the original set of polynomials. We will also see that the portion of fields with more than one corresponding minimal pair (B, C) is insignificant, which is essential. Finishing with a computational consideration of the problems and an investigation of the speed of the convergence of the limit we are considering in Chapter 5, we conclude the discussion of the theory. Important algorithms are introduced and the quintessence of the computational results is presented while more detailed information on the computational work

1. Introduction

(such as source codes) and detailed results can be found in Chapter 6 and in the appendix.

Note that, in the computational part of the work, we often draw on built-in functions of the utilized computer algebra system GP/PARI[1] assuming that they work correctly and fast enough for our purposes—which, in one thing, they do not quite do, as we will see. The bottleneck is the isomorphism-test for two given number fields which is unfortunately responsible for the fact that we have only been able to compute results for relatively small values of X. If this test could be improved significantly—in general or for the special case of totally real cubic fields, we would immediately be able to carry out much more comprehensive computations and choose much larger values for X. Since developing a "better" such test would break the mold of this thesis, we have to be content with what is presented in the appendix.

Another way of getting rid of the described bottleneck up to some degree would of course be the utilization of much more CPU time by using faster computers. All computations have been carried out on a usual modern personal computer so that there was no chance to obtain more or better results. More details on the general computational setup will be given in Chapter 5.

---

[1] The original plan of carrying out all computations with the computer algebra system KASH has been overruled after spotting a bug in KASH that we did not want to bother about, namely the irreducibility-test for polynomials, which does not work correctly. Besides that, GP/PARI seems to be (at least for our purposes) several times faster than KASH.

2. Preliminaries

Let us begin with some preparations. We start with the essential definitions and propositions we will build our theory upon although the fundamental algebraic basics are assumed to be well-known (e.g. field extensions, irreducibility of polynomials etc.) As a great reference for essential algebraic concepts (and beyond), we recommend the textbook by Hungerford, 1974. The key elements we will deal with are totally real algebraic number fields of degree 3. From the beginning on, many of the definitions in this and the following chapters are given in a way which is suitable for our special situation where we are looking at cubic fields as algebraic field extensions of the rational numbers $\mathbb{Q}$ of degree 3. Generalizations are usually straightforward and can easily be derived from the more specialized case. However, if a more general formulation than the one needed for our situation is of any interest (for whatever reason), or is as easy to understand as the specialization to our case, this formulation is given (in these cases, either the specialization to our case is trivial or it is given as a remark) or we refer to an appropriate reference. As a solid introduction to the topic of number fields in general, we refer to the textbook by Marcus, 1977.

2.1. The Theory of Number Fields

At first, we give the most important definition of a totally real algebraic number field which will be essential throughout the whole thesis. We proceed with several additional definitions to introduce notions which will be of high importance for the whole discussion of the theory.

Definition 2.1 (Totally real algebraic number field). An *algebraic number field* K (in the following often only called *number field* or even just *field*) is a finite degree field extension of the rational numbers $\mathbb{Q}$. Thus, $\mathbb{Q}$ is contained in K and K can be considered as a finite-dimensional vector space over $\mathbb{Q}$. K is called totally real if for all embeddings of K into the complex numbers $\mathbb{C}$, the image lies inside the real numbers $\mathbb{R}$. Number fields of degree 3 are called *cubic fields*.

For field elements $\theta \in K$ we will now define the trace and norm as well as the notion of conjugates and the characteristic polynomial.

Definition 2.2 (Trace and norm of a field element). Let $(\omega_1, \ldots, \omega_n)$ be an arbitrary basis of K. For each $\theta \in K$ we have a transformation $T_\theta : K \to K$ defined by

$$T_\theta(y) = \theta y.$$

Hence, for every element $\theta \in K$ there is an associated square matrix $A(\theta) = (a_{ij})_{1 \leq i,j \leq n}$ which is defined via

$$\theta \omega_i = \sum_{j=1}^{n} a_{ij} \omega_j, \ i \in \{1, \ldots, n\}.$$

The trace of the element θ, denoted $\text{Tr}(\theta)$, is defined as the trace of $A(\theta)$ and the norm of θ, denoted $\text{N}(\theta)$, is defined as the determinant of $A(\theta)$ where $\text{Tr}(A(\theta))$ is, as usual, the sum of the elements of the main diagonal of $A(\theta)$.

An alternate and, in some cases, more vivid definition can be found in Hungerford, 1974.

Definition 2.3 (Trace and norm of a field element in terms of K-monomorphisms). Let F be a finite dimensional field extension of K and $\bar{K}$ an algebraic closure of K with $F \subseteq \bar{K}$. Let $\sigma_1, \ldots, \sigma_r$ be the distinct K-monomorphisms with $\sigma_j : F \to \bar{K}$ for all $j \in \{1, ..., r\}$. For $\theta \in F$, the trace of θ is the element

$$\text{Tr}(\theta) = [F:K]_i (\sigma_1(\theta) + \sigma_2(\theta) + \cdots + \sigma_r(\theta))$$

and the norm of θ is the element

$$\mathrm{N}(\theta) = (\sigma_1(\theta)\sigma_2(\theta)\cdots\sigma_r(\theta))^{[F:K]_i}$$

where $[F:K]_i$ is the inseparable degree of F over K (see Hungerford, 1974, Definition 6.10).

Definition 2.4 (Conjugates and degree of a field element)**.** Let K be a number field of degree n. For $\theta \in K$ we define its *conjugates* $\theta^{(1)}, \ldots, \theta^{(n)}$ as the n roots of the minimal polynomial $p_\theta(t)$ of θ in K. As usual, $p_\theta \in \mathbb{Q}[t]$ is the monic polynomial of least degree k such that $p_\theta(\theta) = 0$. The number k is called *degree* of the field element θ. We denote the k distinct values among the set of the $\theta^{(j)}$ by $\theta_1, \ldots, \theta_k$.

Definition 2.5 (Characteristic polynomial of a field element)**.** Let K be a number field of degree n. For $\theta \in K$ we define the *characteristic polynomial* of θ as

$$\chi_\theta(t) := \prod_{j=1}^{n}(t - \theta^{(j)})$$

where $\theta^{(j)}$ are the n conjugates of θ.

Remark 2.6. Let k be the degree of a field element θ in a number field $K = \mathbb{Q}(x)$ of degree n. Then $k \mid n$ and an equivalent definition of the trace and the norm of θ can be given in terms of $\theta^{(1)}, \ldots, \theta^{(n)}$ and $\theta_1, \ldots, \theta_k$. With $m := \frac{n}{k}$ we have

$$\mathrm{Tr}(\theta) = \sum_{j=1}^{n} \theta^{(j)} = m \sum_{j=1}^{k} \theta_j$$

for the trace as well as

$$\mathrm{N}(\theta) = \prod_{j=1}^{n} \theta^{(j)} = \left(\prod_{j=1}^{k} \theta_j\right)^m$$

for the norm (see Milne, 2009, Proposition 2.19 and Corollary 2.20). The minimal polynomial p_θ can of course be written as

$$p_\theta(t) = t^k - a_{k-1}t^{k-1} + \cdots + (-1)^k a_0$$

and the characteristic polynomial χ_θ is given by

$$\chi_\theta(t) = \prod_{j=1}^{k}(t - \theta_j)^m$$

so that altogether this leads to the fact that we can use $\operatorname{Tr}(\theta) = -ma_{k-1}$ and $\operatorname{N}(\theta) = a_0^m$, respectively, as expressions for the trace and the norm of θ. In particular, for field elements θ of degree n, we have $\operatorname{Tr}(\theta) = -a_{n-1}$ and $\operatorname{N}(\theta) = a_0$. A root x of the polynomial f which defines the number field K is an example for a field element of degree n.

A fact we will later make use of is given by the following lemma. The lemma is formulated for the case $n = 3$ which we are more interested in than in the general case and for which the necessary calculations are more vivid than for the general case. However, the result can easily be generalized to arbitrary number fields.

Lemma 2.7. *Let K be a totally real number field of degree 3. If $\chi_\theta(t) = t^3 - at^2 + bt - c$ is the characteristic polynomial of $\theta \in K$ then for $\alpha \in \mathbb{Q}$ the characteristic polynomial of $\alpha\theta \in K$ is given by $\chi_{\alpha\theta}(t) = t^3 - \alpha a t^2 + \alpha^2 b t - \alpha^3 c$.*

Proof. For $1 \leq j \leq 3$ let $\theta^{(j)} \in \mathbb{R}$ be the conjugates of θ. By simple computation we obtain

$$\begin{aligned}\chi_\theta(t) &= (t - \theta^{(1)})(t - \theta^{(2)})(t - \theta^{(3)}) \\ &= t^3 \\ &\quad - t^2(\theta^{(1)} + \theta^{(2)} + \theta^{(3)}) \\ &\quad + t(\theta^{(1)}\theta^{(2)} + \theta^{(1)}\theta^{(3)} + \theta^{(2)}\theta^{(3)}) \\ &\quad - \theta^{(1)}\theta^{(2)}\theta^{(3)}\end{aligned}$$

and we have

$$\begin{aligned} a &= \theta^{(1)} + \theta^{(2)} + \theta^{(3)}, \\ b &= \theta^{(1)}\theta^{(2)} + \theta^{(1)}\theta^{(3)} + \theta^{(2)}\theta^{(3)}, \\ c &= \theta^{(1)}\theta^{(2)}\theta^{(3)} \end{aligned}$$

in $\chi_\theta(t) = t^3 - at^2 + bt - c$. On the other hand, the characteristic polynomial of $\alpha\theta$ is given by

$$\begin{aligned} \chi_{\alpha\theta}(t) &= (t - \alpha\theta^{(1)})(t - \alpha\theta^{(2)})(t - \alpha\theta^{(3)}) \\ &= t^3 - \alpha t^2(\theta^{(1)} + \theta^{(2)} + \theta^{(3)}) \\ &\quad + \alpha^2 t(\theta^{(1)}\theta^{(2)} + \theta^{(1)}\theta^{(3)} + \theta^{(2)}\theta^{(3)}) \\ &\quad - \alpha^3 \theta^{(1)}\theta^{(2)}\theta^{(3)} \\ &= t^3 - \alpha a t^2 + \alpha^2 b t - \alpha^3 c \end{aligned}$$

which finishes the proof. $\square$

Another important notion which we will use several times (and which is interesting for almost all kinds of investigations of number fields) is the field discriminant of an algebraic number field.

Definition 2.8 (Field discriminant). Let K be a totally real number field of degree n and let σ_i, $i \in \{1, \ldots, n\}$, be the n embeddings of K in $\mathbb{C}$. For a set $\{\theta_j \in K \mid 1 \leq j \leq n\}$ of n elements[2] of K, we define the discriminant of $(\theta_1, \ldots, \theta_n)$ as

$$\begin{aligned} d((\theta_1, \ldots, \theta_n)) &:= \det\left((\sigma_i(\theta_j))_{1 \leq i,j \leq n}\right)^2 \\ &= \det\left((\theta_j^{(i)})_{1 \leq i,j \leq n}\right)^2. \end{aligned}$$

The discriminant of a $\mathbb{Z}$-basis of the ring of integers of K (called integral basis), denoted $\mathcal{O}_K$, is called *field discriminant* (or just *discriminant*) of K and is denoted $d(K)$.

---

[2] Note that here the θ_j represent independent field elements of K and not distinct conjugates of one single field element θ as before.

2. Preliminaries

The notion of successive minima has been introduced by Minkowski for lattices Λ in general (see Pohst and Zassenhaus, 1989). We give the definition for the slightly specialized case $\Lambda = \mathcal{O}_K$ which is suitable for our theory. Note that $\text{Tr}(x^2)$ is a positive definite quadratic form on $\mathcal{O}_K$. We prepend the definition of a reduced basis.

Definition 2.9 (Reduced basis). Define a partial ordering on the set of all bases of $\mathcal{O}_K$ via

$$(\omega_1, \ldots, \omega_n) < (\omega_1', \ldots, \omega_n')$$
$$\Leftrightarrow \exists j \in \{1, \ldots, n\} \forall i \in \{1, \ldots, j-1\} : ||\omega_i|| = ||\omega_i'|| \wedge ||\omega_j|| < ||\omega_j'||.$$

A minimal basis of $\mathcal{O}_K$ with respect to $<$ is called *(Minkowski) reduced basis* of $\mathcal{O}_K$.

Definition 2.10 (Successive minima of a number field). Let $(\omega_1 = 1, \omega_2, \ldots, \omega_n)$ be a reduced basis of K. Then the *successive minima* of K are given by

$$M_i(K) = \text{Tr}(\omega_i^2)$$

for $1 \leq i \leq n$. In particular this means $M_2(K) = \text{Tr}(x^2)$ if the "right" root x of a polynomial f is chosen so that $K = \mathbb{Q}(x)$. In general we have $M_2(K) \leq \text{Tr}(x^2)$ while $M_1(K) = n$ is always true due to the properties of arithmetic an geometric mean.

Let now K be a number field over $\mathbb{Q}$ with degree 3 and discriminant $d(K)$. Let $M_2(K)$ be the second successive minimum of K and let $X \in \mathbb{N}$. K has a defining monic irreducible polynomial (over $\mathbb{Q}$) of the form

$$f(t) = t^3 + at^2 + bt + c$$

which means that $K = \mathbb{Q}(x)$ for a root x of f. We require f to have three distinct real roots so that K is totally real. We then have $d(f) > 0$ for the polynomial discriminant and, consequently, also $d(K) > 0$.

If x is a root of $f(t) = t^3 + at^2 + bt + c$ then, by the definition of the trace, we obtain that $\text{Tr}(x)$ is related to the polynomial coefficients a and b via

$$a^2 = \text{Tr}(x)^2 = \text{Tr}(x^2) + 2b$$

where the first equality is clear and the second equation follows from Vieta's formulas. In the following we will require $\text{Tr}(x^2) = a^2 - 2b \leq X$ which implies $M_2(K) \leq X$.

Definition 2.11. From now on, let $F(X)$ be the set of isomorphism classes of totally real cubic fields K with $M_2(K) \leq X$.

2.2. Class Field Theory

Notions like the class number and the regulator of an algebraic number field will occur in the course of this thesis and, for the sake of completeness, we give the relevant definitions without elaborating on details right now.

Definition 2.12 (Class group). Let K be a number field and $\mathcal{O}_K$ the ring of integers of K. Two fractional ideals I and J of K are said to be equivalent if there is $\alpha \in K^\times$ with $J = \alpha I$. The set of equivalence classes is called the class group of K or of $\mathcal{O}_K$ and is denoted $Cl(K)$.

The class group $Cl(K)$ is finite (and abelian) and its cardinality is called the *class number*, denoted $h(K)$.

For the definition of the regulator of a number field, we first have to introduce the notion of the logarithmic embedding.

Definition 2.13 (Logarithmic embedding). The *logarithmic embedding* of $K^\times$ in $\mathbb{R}^{r_1+r_2}$ is the mapping $L(\theta)$ which is defined by

$$K^\times \to \mathbb{R}^{r_1+r_2}$$
$$\theta \mapsto (\ln|\sigma_1(\theta)|, \ldots, \ln|\sigma_{r_1}(\theta)|, 2\ln|\sigma_{r_1+1}(\theta)|, \ldots, 2\ln|\sigma_{r_1+r_2}(\theta)|)$$

where r_1 is the number of real embeddings[3] of K (or $K^\times$) and $2r_2$ is the number of complex embeddings of K (or $K^\times$).

Remark 2.14. Definition 2.13 simplifies to

$$K^\times \to \mathbb{R}^3$$
$$\theta \mapsto (\ln|\sigma_1(\theta)|, \ln|\sigma_2(x)|, \ln|\sigma_3(\theta)|)$$

in the case of a totally real cubic number field, since then $r_1 = 3$ and $r_2 = 0$.

It can be shown that the image of $K^\times$ under the mapping L is a lattice in the hyperplane $\sum_{1 \leq i \leq r_1 + r_2} \theta_i = 0$ of $\mathbb{R}^{r_1 + r_2}$ in general or in the hyperplane $\theta_1 + \theta_2 + \theta_3 = 0$ of $\mathbb{R}^3$ in the case of totally real cubic fields (see Cohen, 1993, Chapter 4.9.2). With these preparations, we can now define the regulator of a number field.

Definition 2.15 (Regulator). The volume of the above-mentioned lattice (the absolute value of the determinant of any $\mathbb{Z}$-basis of the lattice) is called the *regulator* of K, denoted $R(K)$.

The computation of regulators, as well as the computation of class numbers, is a main topic in computational algebraic number theory. Whenever we need such values for computations in the following, we will not concentrate on theoretical aspects such as how to compute them as fast as possible, but draw on functionalities and routines built into the utilized computer algebra system. In particular, for our special case where only totally real cubic number fields are of interest, the values $h(K)$ and $R(K)$ can be calculated with GP/PARI in practically no time if the defining polynomial for K is given (see source codes in the appendix).

---

[3] "Real embeddings" of K means embeddings of K into the complex numbers $\mathbb{C}$ whose images lie inside the real numbers $\mathbb{R}$, and "complex embeddings" analogously means embeddings of K into the complex numbers $\mathbb{C}$ whose images lie inside $\mathbb{C} \setminus \mathbb{R}$.

2.3. Additional Preparations

We will later refer to the Riemann zeta function and the Dedekind zeta function which are defined as follows.

Definition 2.16 (Riemann zeta function). For $s \in \mathbb{C}$ with $\Re(s) > 1$ the power series

$$\zeta(s) := \sum_{n=1}^{\infty} n^{-s} \qquad (2.1)$$

is called the Riemann zeta function. An equivalent expression of 2.1 is

$$\zeta(s) = \prod_{p \in \mathbb{P}} \left(1 - \frac{1}{p^s}\right)^{-1} \qquad (2.2)$$

where $\mathbb{P}$ is the set of all prime numbers.

Definition 2.17 (Dedekind zeta function). The Dedekind zeta function of a number field K is defined as

$$\zeta_K(s) := \sum_{I \subseteq \mathcal{O}_K} |\mathcal{O}_K/I|^{-s} \qquad (2.3)$$

where I runs over all non-zero ideals of the ring of integers of K, $\mathcal{O}_K$ and $\mathcal{O}_K/I$ is a quotient ring.

3. A Finite Set of Polynomials

In this chapter, we will build the finite set $P(X)$ of polynomials which is defined only by few conditions. We will work out the bounds for the coefficients of the linear and the constant term of the polynomials in this set, since the coefficient for the cubic term is 1 for monic polynomials and the coefficient of the quadratic term is bounded by assumption. We will show how the cardinality of the set $P(X)$ can be computed easily and we will think about the errors that might arise in the chosen procedure. We will see that, as a matter of fact, there *are* errors but that these errors can be neglected since they do not have an effect asymptotically because their order is small enough.

3.1. Preparations

From the above, we have the necessary preparations and definitions to start the discussion of our theory. We will first derive a finite set of polynomials of the form $f(t) = t^3 + at^2 + bt + c$ only from the two conditions $a \in \{0, 1, 2\}$ for the polynomial coefficient a of the quadratic term and $\text{Tr}(x^2) = a^2 - 2b \leq X$ for a root x of f. From this, we obtain bounds for the polynomial coefficients b and c by simple calculation.

Using the two conditions mentioned above, the fact that $\text{Tr}(x^2) > 0$ (since the field defined by f is required to be totally real), and the equation

$$d(f) = a^2b^2 - 4b^3 - 4a^3c + 18abc - 27c^2$$

for the discriminant of a polynomial of the form $t^3 + at^2 + bt + c$ (which follows from Vieta's formulas), we obtain

$$\frac{a^2 - X}{2} \leq b < \frac{a^2}{2}$$

as well as

$$-\frac{1}{27}\left(2(a^2 - 3b)^{3/2} + (2a^3 - 9ab)\right) < c < \frac{1}{27}\left(2(a^2 - 3b)^{3/2} - (2a^3 - 9ab)\right)$$

as bounds for the coefficients b and c, respectively. Note that $a^2 - 3b > 0$. This can be seen as follows: since the trace form $\mathrm{Tr}(\cdot)$ is positive definite (because the field K defined by f is requested to be totally real), we get

$$\begin{aligned} 0 &< n\mathrm{Tr}\left(\left(x - \frac{\mathrm{Tr}(x)}{n}\right)^2\right) \\ &= n\mathrm{Tr}(x^2) - 2\mathrm{Tr}(x)^2 + \mathrm{Tr}(x)^2 \\ &= n\mathrm{Tr}(x^2) - \mathrm{Tr}(x)^2. \end{aligned}$$

This implies

$$\begin{aligned} 0 &< n(a^2 - 2b) - a^2 \\ &= (n-1)a^2 - 2nb \end{aligned}$$

which becomes $2a^2 - 6b = 2(a^2 - 3b) > 0$ in the case $n = 3$. In particular, this gives a slightly refined upper bound for b, namely $b < \frac{a^2}{3}$ and we obtain, due to the fact that b is an integer, that $b \leq \lfloor \frac{a^2}{3} \rfloor$. Note that for reasons of consistency it must also hold that $2(a^2 - 3b)^{3/2} - (2a^3 - 9ab) \geq 0$ and $2(a^2 - 3b)^{3/2} + (2a^3 - 9ab) \geq 0$. However, this can also be seen easily by considering the three cases for a separately.

For $a = 0$ we have

$$\begin{aligned} 2(a^2 - 3b)^{3/2} - (2a^3 - 9ab) &= 6\sqrt{3}(-b)^{3/2} \\ &\geq 0 \end{aligned}$$

3. A Finite Set of Polynomials

as well as

$$2(a^2 - 3b)^{3/2} + (2a^3 - 9ab) = 6\sqrt{3}(-b)^{3/2}$$
$$\geq 0$$

since for $a = 0$ we know that $-b \geq 0$.

For $a = 1$ we have

$$2(a^2 - 3b)^{3/2} - (2a^3 - 9ab) = 2(1 - 3b)^{3/2} + 9b - 2$$
$$\geq 0$$

because the function $f(b) = 2(1-3b)^{3/2} + 9b - 2$ has a local minimum at 0. This is due to $f'(b) = 9 - 9\sqrt{1-3b}$, hence $f'(0) = 0$ and $f''(b) = 27/(2\sqrt{1-3b})$, hence $f''(0) > 0$. It is clear from the form of f that there exists only one minimum, therefore 0 is a global minimum. Note also that all possible choices for b lie within the domain of f. Furthermore we have

$$2(a^2 - 3b)^{3/2} + (2a^3 - 9ab) = 2(1 - 3b)^{3/2} - 9b + 2$$
$$\geq 2(1 - 3b)^{3/2} + 9b - 2$$
$$\geq 0$$

since for $a = 1$ we have $b \leq 0$ while the last inequality holds for $b \leq 1/3$.

For $a = 2$ we have

$$2(a^2 - 3b)^{3/2} - (2a^3 - 9ab) = 2(4 - 3b)^{3/2} + 2(9b - 8)$$
$$\geq 0$$

because the function $f(b) = 2(4 - 3b)^{3/2} + 2(9b - 8)$ has a local minimum at 0. This is due to $f'(b) = 18 - 9\sqrt{4-3b}$, hence $f'(0) = 0$ and $f''(b) = 27/(2\sqrt{4-3b})$, hence $f''(0) > 0$. It is clear from the type of f that there exists only one minimum, therefore 0 is a global minimum. Note again that

3. A Finite Set of Polynomials

all possible choices for b lie within the domain of f. Furthermore we have, for similar reasons as in the case $a = 1$, that

$$2(a^2 - 3b)^{3/2} + (2a^3 - 9ab) = 2(4 - 3b)^{3/2} - 2(9b - 8)$$
$$\geq 0.$$

Altogether this enables us to think about how large the set $P(X)$ of all polynomials in question due to the above restrictions and bounds is. The following proposition sums up what we have just seen.

Proposition 3.1. *For a polynomial $f(t) = t^3 + at^2 + bt + c$, the conditions $0 \leq a \leq 2$, $0 < a^2 - 2b$, and $d(f) > 0$ are equivalent to the following bounds on the coefficients:*

$$\frac{a^2 - X}{2} \leq b < \frac{a^2}{2} \quad \text{and}$$

$$-\frac{1}{27}\left(2(a^2 - 3b)^{3/2} + (2a^3 - 9ab)\right) < c < \frac{1}{27}\left(2(a^2 - 3b)^{3/2} - (2a^3 - 9ab)\right)$$

and the set $P(X)$ of all irreducible polynomials satisfying these conditions is finite.

3.2. Counting Polynomials

Let now $P(X)$ be the finite set of polynomials whose coefficients satisfy the above conditions and let us compute $|P(X)|$. The canonical way to do this is of course to consider double sums

$$|P(X)| \sim \frac{1}{27} \sum_{a=0}^{2} \sum_{b=\lceil(a^2-X)/2\rceil}^{\lfloor a^2/3\rfloor} 4(a^2 - 3b)^{3/2} \tag{3.1}$$

3. A Finite Set of Polynomials

where we can express the inner sum as an appropriate integral so that we get

$$|P(X)| \sim \frac{1}{27} \sum_{a=0}^{2} \int_{(a^2-X)/2}^{\lfloor a^2/3 \rfloor} 4(a^2 - 3b)^{3/2} db \qquad (3.2)$$

$$= \frac{\sqrt{6} X^{5/2}}{45}$$

$$+ \frac{\sqrt{2}((3X-1)^{5/2} - 4\sqrt{2})}{405}$$

$$+ \frac{\sqrt{2}((3X-4)^{5/2} - 4\sqrt{2})}{405}$$

where $\sim$ in (3.1) and (3.2) means that we have to think about the error that arises by changing from summation to integration and by using the floor and ceiling function in the bounds for summation and dropping the ceiling function again in the bounds for integration. We do this to obtain nicer results and since computations for large values of X suggest that

$$\lim_{X \to \infty} |P(X)| \cdot X^{-5/2} = \frac{\sqrt{6}}{15} \qquad (3.3)$$

in any case (which will become clear later), it follows that we are allowed to do this and that the number of polynomials to be considered is of the same order as $X^{5/2}$. See Figure 1 and Table 1 to get an idea of how fast the convergence in (3.3) actually is. The emerging errors will be treated in detail later.

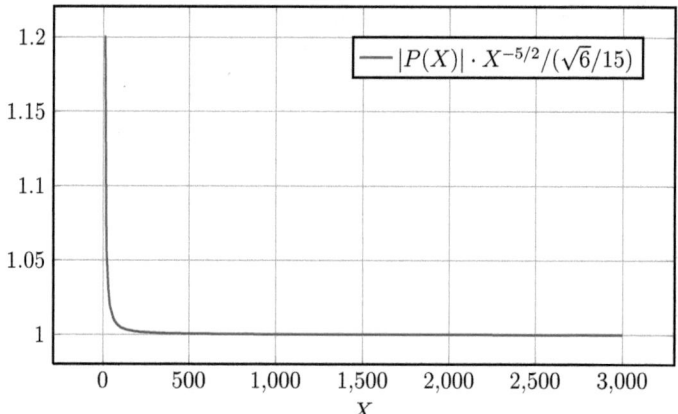

Figure 1.: Value of the fraction $\frac{|P(X)| \cdot X^{-5/2}}{\sqrt{6}/15}$

X	Value
10	1.20062484
100	1.00520935
1000	1.00031898
2000	1.00015121
3000	1.00009871

Table 1.: Value of the fraction $\frac{|P(X)| \cdot X^{-5/2}}{\sqrt{6}/15}$

We know that $|F(X)| \sim X^{5/2}$ from ("almost-") equality (3.2), therefore the above computations imply that the number of reducible polynomials (which do not generate number fields) in the finite set $P(X)$ is negligible. In fact, it is of the order $\mathcal{O}(X^{3/2})$ as the following proposition shows.

Proposition 3.2. *The number of reducible polynomials among all polynomials in the set $P(X)$ is of the order $\mathcal{O}(X^{3/2})$.*

3. A Finite Set of Polynomials

Proof. Let $f(t) = t^3 + at^2 + bt + c$ and x a root of f. We then have $x^2 \leq \text{Tr}(x^2) \leq X$ which implies

$$|x| \leq X^{1/2}.$$

For general polynomials $f(t) = t^3 + at^2 + bt + c$, applying polynomial long division (a division algorithm for polynomials, cf. Hungerford, 1974, Chapter III.6, Theorem 6.2.) by $(t - x)$, we obtain that

$$\begin{aligned} f(t) &= (t-x)(t^2 + (a+x)t + b + ax + x^2) \\ &\quad + c + bx + ax^2 + x^3 \end{aligned}$$

where $t^2 + (a+x)t + b + ax + x^2$ is the quotient of $t^3 + at^2 + bt + c$ divided by $(t - x)$ and $c + bx + ax^2 + x^3$ is the remainder of the division. If f is reducible then for a suitable root x of f we know that $(t - x)$ is a factor of f. The division of f by this factor yields of course a remainder which is equal to zero. Another factor of f is hence given by

$$t^2 + (a+x)t + b + ax + x^2.$$

Setting $d := b + ax + x^2$ we can therefore write

$$\begin{aligned} f(t) &= (t-x)(t^2 + (a+x)t + d) & (3.4) \\ &= t^3 + at^2 + (d - x(a+x))t - dx \end{aligned}$$

where the discriminant of $t^2 + (a+x)t + d$ is (due to Vieta's formulas) given by $(a+x)^2 - 4d > 0$ which is equivalent to $d < (a+x)^2/4$. This, together with the fact that

$$\begin{aligned} \sum_{k=1}^{X^{1/2}} k^2 &= \frac{1}{6}(X^{1/2})(X^{1/2}+1)(2X^{1/2}+1) & (3.5) \\ &= \mathcal{O}(X^{3/2}) \end{aligned}$$

yields that there are at most $\mathcal{O}(X^{3/2})$ reducible polynomials (note that we still demand that $a \in \{0, 1, 2\}$ which gives the necessary limitation for the number of coefficients $(a + x)$ in (3.4)). Of course, (3.5) also works with $\lceil X^{1/2} \rceil$ as upper summation bound if $X^{1/2}$ is not an integer. The ceiling function then has to be added to the right side of (3.5) whence the order of the value of the sum remains the same. □

In the next section we will estimate the errors which we have already thought about briefly in order to see that they can fortunately be neglected.

3.3. Estimation of Errors

Let us first consider the easier-to-handle source of errors in the above formulas. The first error that comes up is because we use the floor and ceiling functions in the summation bounds in (3.1) which gives only an error of constant order $\mathcal{O}(1)$ for any possible choice of a. This error transfers to the expression via integral in (3.2) where we drop the ceiling function in the lower bound (we do that to avoid nasty results when computing the integral). However, its order clearly stays the same, i.e. $\mathcal{O}(1)$. A further error might arise in the inner sum. In every summation step we can miscount one integer which, together with the fact that we have kX summation steps (for a constant k), leads to an error of the order $\mathcal{O}(X)$.

The slightly harder to handle and, as we will see, more significant (but still negligible) source of errors is the transition from the inner sum to the integral, which can fortunately be investigated by applying the Euler-Maclaurin formula (also known as Euler's summation formula) which we will present with a proof from Apostol, 1998.

3. A Finite Set of Polynomials

Theorem 3.3 (Euler-Maclaurin formula). *Let f be a function that has a continuous derivative f' on the closed interval $[y, x]$ where $0 < y < x$. Then the equality*

$$\sum_{y<n\leq x} f(n) = \int_y^x f(t)dt + \int_y^x (t - \lfloor t \rfloor) f'(t) dt + f(x)(\lfloor x \rfloor - x) - f(y)(\lfloor y \rfloor - y)$$

holds.

Proof. Let $\bar{y} = \lfloor y \rfloor$ and $\bar{x} = \lfloor x \rfloor$. For $n \in \mathbb{Z}$ and $n - 1 \in [y, x]$ we have

$$\begin{aligned}\int_{n-1}^n \lfloor t \rfloor f'(t) dt &= \int_{n-1}^n (n-1) f'(t) dt \\ &= (n-1)(f(n) - f(n-1)) \\ &= (nf(n) - (n-1)f(n-1)) - f(n).\end{aligned}$$

Summing from $\bar{y} + 1$ to $\bar{x}$ we obtain

$$\begin{aligned}\int_{\bar{y}}^{\bar{x}} \lfloor t \rfloor f'(t) dt &= \sum_{n=\bar{y}+1}^{\bar{x}} (nf(n) - (n-1)f(n-1)) - \sum_{y<n\leq x} f(n) \\ &= \bar{x} f(\bar{x}) - \bar{y} f(\bar{y}) - \sum_{y<n\leq x} f(n)\end{aligned}$$

and therefore

$$\begin{aligned}\sum_{y<n\leq x} f(n) &= -\int_{\bar{y}}^{\bar{x}} \lfloor t \rfloor f'(t) dt + \bar{x} f(\bar{x}) - \bar{y} f(\bar{y}) \qquad (3.6) \\ &= -\int_y^x \lfloor t \rfloor f'(t) dt + \bar{x} f(x) - \bar{y} f(y).\end{aligned}$$

With integration by parts we get

$$\int_y^x f(t) dt = xf(x) - yf(y) - \int_y^x tf'(t) dt. \qquad (3.7)$$

Combining (3.6) with (3.7) we obtain the claim. $\square$

The conditions of the theorem are met since we are considering the function

$$f(b) = 4(a^2 - 3b)^{3/2}$$

whose first derivative
$$f'(b) = -18\sqrt{a^2 - 3b}$$
is continuous on the closed interval $[(a^2 - X)/2, a^2/3]$ for $a \in \{0, 1, 2\}$. Consequently, we can apply Euler's formula to our inner sum which then yields

$$\sum_{b=\lceil(a^2-X)/2\rceil}^{\lfloor a^2/3 \rfloor} 4(a^2 - 3b)^{3/2}$$

$$\begin{aligned} = &\int_{(a^2-X)/2}^{\lfloor a^2/3 \rfloor} (4(a^2 - 3b)^{3/2}) db \\ + &\int_{(a^2-X)/2}^{\lfloor a^2/3 \rfloor} (b - \lfloor b \rfloor) \cdot (-18\sqrt{a^2 - 3b}) db \\ - &\sqrt{2}(3X - a^2)^{3/2}(\lfloor (a^2 - X)/2 \rfloor - (a^2 - X)/2) \\ + &\mathcal{O}(1) \end{aligned}$$

with the term $\mathcal{O}(1)$ coming from the omitted ceiling functions in the lower integration bound (which we do again to avoid nasty results when computing the integral). The value in the last brackets in the second to last line is either 0 or $-\frac{1}{2}$, hence the value of the last term is either 0 or $\frac{1}{2}\sqrt{2}(3X - a^2)^{3/2}$, leading to an error which is of the order $\mathcal{O}(X^{3/2})$. It remains for us to have a closer look at the second term (the third to last line). Note that

$$\left| \int_{(a^2-X)/2}^{\lfloor a^2/3 \rfloor} (b - \lfloor b \rfloor)(-18\sqrt{a^2 - 3b}) db \right| \leq \left| \int_{(a^2-X)/2}^{\lfloor a^2/3 \rfloor} (-18\sqrt{a^2 - 3b}) db \right|$$

since $(b - \lfloor b \rfloor) \in [0, 1)$. We treat the cases $a = 0$, $a = 1$, and $a = 2$ for the integral on the right hand separately.

For $a = 0$ we obtain

$$\left| \int_{-X/2}^{0} -18\sqrt{-3b} db \right| = \sqrt{54} |X|^{3/2}$$

which gives an error of the order $\mathcal{O}(X^{3/2})$.

For $a = 1$ we obtain

$$\left| \int_{(1-X)/2}^{0} -18\sqrt{1-3b}\,db \right| = \sqrt{2}\left|(3X-1)^{3/2} - 2\sqrt{2}\right|$$

and the error is again of the order $\mathcal{O}(X^{3/2})$.

Finally, for $a = 2$ we obtain

$$\left| \int_{(4-X)/2}^{1} -18\sqrt{4-3b}\,db \right| = \sqrt{2}\left|(3X-4)^{3/2} - 2\sqrt{2}\right|$$

and the error is once more of the order $\mathcal{O}(X^{3/2})$.

Altogether, with the number of reducible polynomials we have to deal with being of no importance and the fact that the largest error that occurs when computing $|P(X)|$ is of the order $\mathcal{O}(X^{3/2})$, we have proved the following important result.

Theorem 3.4. *For the size of the set $P(X)$ we have*

$$|P(X)| = \frac{\sqrt{6}}{15}X^{5/2} + \mathcal{O}(X^{3/2}).$$

We have seen that the error that occurs by switching from sums to integrals is always of the order $\mathcal{O}(X^{3/2})$. Hence the "greatest" error we have to deal with is of the order $\mathcal{O}(X^{3/2})$ and so is the overall error. Since furthermore $X^{3/2} \ll X^{5/2}$ for large X, we have proved that using integrals instead of sums for counting polynomials in the set $P(X)$ is suitable, particularly because later we are interested in the results for large values of X.

4. Parametrization of the Polynomials

We now introduce another possibility of expressing the polynomials considered above which will lead us eventually to the proof of the limit

$$\lim_{X \to \infty} \frac{|P_\kappa(X)|}{|F(X)|} = 1$$

where the size of $P_\kappa(X)$ is equal to the size of the set $P(X)$ modulo a relation $\sim_P$ which will be explained in the following.

If a, b, and c are the coefficients of a polynomial $f(t) = t^3 + at^2 + bt + c$ and x is a root of f that lies in the number field K defined by f then we define a parametrization of f as follows.

Definition 4.1. If f is a defining polynomial for the field K then a parametrization $(B, C) = (B(x), C(x))$ of f is given by

$$\begin{aligned}
B(x) &:= \frac{3}{2}\mathrm{Tr}\left(\left(x - \frac{\mathrm{Tr}(x)}{3}\right)^2\right) \\
&= \frac{3}{2}\mathrm{Tr}\left(\left(x - \frac{a}{3}\right)^2\right) \\
&= \frac{3}{2}\mathrm{Tr}\left(x^2 - \frac{2}{3}ax + \frac{a^2}{9}\right) \\
&= \frac{3}{2}\left(a^2 - 2b - \frac{2}{3}a^2 + \frac{1}{9}a^2\right) \\
&= a^2 - 3b
\end{aligned}$$

4. Parametrization of the Polynomials

and

$$\begin{aligned} C(x) &:= 27\mathrm{N}\left(\frac{a}{3} - x\right) \\ &= -2a^3 + 9ab - 27c \\ &= a^3 - 3aB - 27c. \end{aligned}$$

While the equalities for B are clear due to the definition and computation rules for the trace, the second equality for C can be seen as follows: by definition we know that $\mathrm{N}(\frac{a}{3} - x)$ is the constant term of the polynomial

$$\begin{aligned} &(\frac{a}{3} - x)^3 - a(\frac{a}{3} - x)^2 + b(\frac{a}{3} - x) - c \\ &= -\frac{1}{27}(27x^3 + 9x(3b - a^2) + 2a^3 - 9ab + 27c) \end{aligned}$$

in x which is $-\frac{1}{27}(2a^3 - 9ab + 27c)$. We see that B and C are obviously invariant under the translation $x \mapsto x + \beta$ with $\beta \in \mathbb{Q}$ and, of course, we have $B > 0$ and $C \neq 0$ from the definition of the trace and the norm, respectively. If a, b, and c are integers then we have the following result.

Corollary 4.2. *For integers a, b, and c with $0 \leq a \leq 2$ we have $a \in \mathbb{Z}$ with $B \equiv a^2 \pmod{3}$ and $C \equiv a^3 - 3aB \pmod{27}$ if and only if one of the three following pairs of congruences holds. We either have*

$$B \equiv 0 \pmod{3} \text{ and } C \equiv 0 \pmod{27} \tag{4.1}$$

or

$$B \equiv 1 \pmod{3} \text{ and } C \equiv 1 - 3B \pmod{27} \tag{4.2}$$

or

$$B \equiv 1 \pmod{3} \text{ and } C \equiv 3B - 1 \pmod{27}. \tag{4.3}$$

Remark 4.3. From now on we are only considering irreducible polynomials $f(t)$ and we have seen the reasons for this in the previous chapter: we have introduced the notions of the trace and the norm in a way that only works with

4. Parametrization of the Polynomials

irreducible polynomials. Furthermore, the portion of reducible polynomials (which do not generate number fields) is asymptotically equal to zero.

Note that the values of B and C only depend on a (mod 3). A pair (B, C) is related uniquely (i.e. there is a $1-1$ correspondence) to a polynomial in the set $P(X)$ and we will now derive bounds for B and C so that we can count these pairs instead of polynomials. However, the number of pairs is not what we actually want to count—we have to "cut back" this value in a certain way and the following steps are only a necessary preparation for this. Recall that we have earlier computed

$$\frac{a^2 - X}{2} \leq b < \frac{a^2}{3}$$

as bounds for b. With $b = \frac{a^2 - B}{3}$ and $B > 0$ we therefore get

$$1 \leq B \leq \frac{3X - a^2}{2}$$

and by accepting a negligible error (recall that we are interested in the results for large values of X) we can choose

$$1 \leq B \leq \frac{3X}{2}$$

where the 1 as lower bound comes from the fact that B is an integer. A similar computation works for the bounds for C. Recall the bounds for c

$$-\frac{1}{27}\left(2(a^2 - 3b)^{3/2} + (2a^3 - 9ab)\right) < c < \frac{1}{27}\left(2(a^2 - 3b)^{3/2} - (2a^3 - 9ab)\right).$$

Substituting again b with $\frac{a^2 - B}{3}$ and c with $-\frac{3aB + C - a^3}{27}$ we obtain

$$-2B^{3/2} < C < 2B^{3/2} \tag{4.4}$$

and we are now able to write (B, C) as a substitute for $t^3 + at^2 + bt + c$ while the number of pairs counted is equal to the number of polynomials counted (up

4. Parametrization of the Polynomials

to possible errors as stated above). The following proposition sums up what we have just explained and the statement will become clear in the following.

Proposition 4.4. *With the conditions from Corollary 4.2 and the bounds*

$$1 \leq B \leq \frac{3X}{2}$$

for the possible values of B as well as

$$-2B^{3/2} < C < 2B^{3/2}$$

for the possible values of C, we count a number of pairs (B,C) that is asymptotically equal to the counted number of (irreducible[4]) polynomials defined by coefficients a, b, and c where a, b, c and (B,C) are related as described above.

We think about how to count pairs (B,C) and, in one shot, work out how the number of those pairs is related to the number of polynomials.

To count pairs (B,C) within bounds defined by X, we can again consider sums of the form

$$\sum_{B=1}^{\lfloor 3X/2 \rfloor} 4B^{3/2}$$

with the conditions for B and for the value in the sum (related to C) as above. These conditions, however, make further calculations unnecessarily hard so we choose a quite simple way to get rid of them. Suppose we calculate the value of the above sum without paying attention to the modulo-constraints.

Of course, $B \equiv 0 \pmod 3$ is the case for $\frac{1}{3}$ of all values that B can attain in the given interval. Similarly, for the value in the sum, the corresponding modulo-constraint, $C \equiv 0 \pmod{27}$ which holds for $\frac{1}{27}$ of all cases, is relevant. We therefore have to multiply the value obtained so far by $\frac{1}{3} \cdot \frac{1}{27} = \frac{1}{81}$. Because the same argument holds for the other two pairs of equivalences, we have three possible pairs of equivalences, each occurring in $\frac{1}{81}$ of all cases. Altogether that

---

[4] Note again that the portion of reducible polynomials in $P(X)$ is asymptotically equal to zero.

means that we can calculate the value of the sum without paying attention to the modulo-conditions at first and subsequently multiply the result by $\frac{1}{81} \cdot 3 = \frac{1}{27}$ to be back on track. The number of pairs (B, C) can hence be computed by evaluating

$$\frac{1}{27} \sum_{B=1}^{\lfloor 3X/2 \rfloor} 4B^{3/2} \qquad (4.5)$$

without having to deal with further constraints on the variables.

Once more, we can try to change this sum to an integral as we did above which gives us the advantage of being able to compute the number of pairs effectively and fast but which again compels us to think about the errors that arise by doing that. Writing the sum as an integral, we drop the ceiling function in the upper summation bound and choose 0 instead of 1 as lower bound in order to obtain a slightly nicer result when computing the integral (we can do this since the error arising from that is only of the order $\mathcal{O}(1)$) and therefore get

$$\frac{1}{27} \int_1^{\lfloor 3X/2 \rfloor} 4B^{3/2} dB \sim \frac{1}{27} \int_0^{3X/2} 4B^{3/2} dB$$
$$= \frac{\sqrt{6}}{15} X^{5/2} \qquad (4.6)$$

for the number of polynomials. Note that this is exactly the same result as we have seen above (asymptotically) which proves Proposition 4.4.

Altogether, what we have seen so far shows that we can work with pairs (B, C) in the following instead of polynomials defined by their coefficients a, b, and c since the set of pairs is of the same size as the set of polynomials (asymptotically and accepting the discussed errors). Of course there are more polynomials, and hence more pairs (B, C) than isomorphism classes of fields generated by elements of these sets. There are some elements in these sets that can easily be identified as defining element for a number field that has already been defined by another element while it is harder to decide for other elements. The target in the following will therefore be to work out how to "cut back" the set of pairs (B, C) so that each number field which is defined by an element of the set is only represented once in the set. With the results of doing that we will then be able to see that we actually have

$$\lim_{X \to \infty} \frac{|P_\kappa(X)|}{|F(X)|} = 1$$

where $P_\kappa(X)$ is the "cut back" set of pairs (B, C).

4.1. The Minimality of a Pair (B, C)

We will now introduce the notion of p-minimality and minimality of a pair (B, C). This, together with some computations and further thoughts, will eventually cut back the set of pairs (B, C) in a suitable way. In the following we will require the entries B and C of a pair to meet the following conditions:

Definition 4.5 (p-minimality and minimality). For a prime $p \neq 3$, the pair (B, C) is called p-*minimal* if $p^2 \mid B$ and $p^3 \mid C$ does not hold. The pair (B, C) is called 3-*minimal* if the pair $(\tilde{B}, \tilde{C}) = (\frac{B}{9}, \frac{C}{27})$ is not related to a polynomial in the set $P(X)$. (B, C) is called *minimal* if it is p-minimal for all $p \in \mathbb{P}$.

Remark 4.6. 3-minimality is given if the following does not hold:

$$3^3 \mid B,$$
$$3^6 \mid C,$$
$$B = 9\tilde{B} \text{ with } \tilde{B} \equiv 1 \pmod{3},$$
$$C = 27\tilde{C} \text{ with } \tilde{C} \equiv \pm(1 - 3\tilde{B}) \pmod{27}.$$

We only want to consider those pairs (B, C) for which we have $B(\alpha x + \beta) \geq B(x)$ with $\alpha, \beta \in \mathbb{Q}$. We have already seen that B and C are invariant under the translation $x \mapsto x + \beta$ and from the definitions of B and C we will derive easily that if $\alpha x + \beta$ is an algebraic integer then $B(\alpha x + \beta) = \alpha^2 B(x)$ and $C(\alpha x + \beta) = \alpha^3 C(x)$. Altogether that means that we have to exclude $\frac{1}{p}$ for prime numbers p as values for α.

4. Parametrization of the Polynomials

Lemma 4.7. *For $\alpha x + \beta$ with $\alpha, \beta \in \mathbb{Q}$ we have*

$$\mathrm{Tr}(\alpha x + \beta) = \alpha^2 \mathrm{Tr}(x)$$

and

$$\mathrm{N}(\alpha x + \beta) = \alpha^3 \mathrm{N}(x).$$

Proof. We know that the values $B(x)$ and $C(x)$ above are invariant under the translation $x \mapsto x + \beta$. We can therefore consider the case $\beta = 0$ without loss of generality. We know from Lemma 2.7 that if $t^3 - at^2 + bt - c$ is the characteristic polynomial of x then the characteristic polynomial of αx is given by

$$t^3 - \alpha a t^2 + \alpha^2 b t - \alpha^3 c$$

and hence we obtain for the equality regarding the trace:

$$\begin{aligned}
B(\alpha x) &= \frac{3}{2}\mathrm{Tr}\left(\left(\alpha x - \frac{\mathrm{Tr}(\alpha x)}{3}\right)^2\right) \\
&= \frac{3}{2}\mathrm{Tr}\left(\left(\alpha x - \frac{\alpha a}{3}\right)^2\right) \\
&= \frac{3}{2}\mathrm{Tr}\left(\alpha^2 x^2 - \frac{2}{3}\alpha^2 a x + \frac{\alpha^2 a^2}{9}\right) \\
&= \alpha^2 \frac{3}{2}\mathrm{Tr}\left(x^2 - \frac{2}{3}ax + \frac{a^2}{9}\right) \\
&= \alpha^2(a^2 - 3b) \\
&= \alpha^2 B(x).
\end{aligned}$$

For the equality regarding the norm we get in a similar way as before that $\mathrm{N}(\frac{\alpha a}{3} - \alpha x)$ is the constant term of the polynomial

$$(\frac{\alpha a}{3} - \alpha x)^3 - \alpha a(\frac{\alpha a}{3} - \alpha x)^2 + \alpha^2 b(\frac{\alpha a}{3} - \alpha x) - \alpha^3 c$$
$$= -\frac{1}{27}(\alpha^3(27x^3 + 9x(3b - a^2) + 2a^3 - 9ab + 27c)$$

4. Parametrization of the Polynomials   33

which is $-\frac{1}{27}\alpha^3(2a^3 - 9ab + 27c)$. This implies $N(\alpha x) = \alpha^3 N(x)$ and together with the invariance under $x \mapsto x + \beta$ this leads to the assumption. □

Proposition 4.8. *For minimal pairs* (B, C) *the inequality*

$$B(\alpha x + \beta) = \alpha^2 B(x) \geq B(x)$$

always holds.

Proof. For $|\alpha| \geq 1$ we have $B(\alpha x + \beta) \geq B(x)$ due to $B(\alpha x + \beta) = \alpha^2 B(x)$. We therefore have to prove that $(B(\alpha x + \beta), C(\alpha x + \beta))$ cannot be the parametrization of a polynomial in the set $P(X)$ when $\alpha = \frac{1}{p}$ for a prime p. Let $p \neq 3$. Then the first condition of Definition 4.5 which says that there is no p such that $p^2 \mid B$ and $p^3 \mid C$ forbids to choose $\alpha = \frac{1}{p}$.

Let now $p = 3$ so that $\alpha = \frac{1}{3}$. If $(\tilde{B}, \tilde{C}) = (\frac{B}{9}, \frac{C}{27})$ then the conditions $3^3 \nmid B$ and $3^6 \nmid C$ obviously exclude the case where $\tilde{B} \equiv 0 \pmod{3}$ and $\tilde{C} \equiv 0 \pmod{27}$ which would imply that $(\tilde{B}, \tilde{C})$ is a pair related to a polynomial in the set $P(X)$ and that $\alpha^2 B$ is an integer, leading to the unfeasible inequality

$$\alpha^2 B(x) < B(x).$$

If $(\tilde{B}, \tilde{C})$ is uniquely related to a polynomial with integer coefficients a, b, c which is in the set $P(X)$ then we know from before that because of $a \in \{0, 1, 2\}$ this is equivalent to

$$\tilde{B} \equiv 0 \pmod{3} \text{ and } \tilde{C} \equiv 0 \pmod{27}$$

or

$$\tilde{B} \equiv 1 \pmod{3} \text{ and } \tilde{C} \equiv 1 - 3\tilde{B} \pmod{27}$$

or

$$\tilde{B} \equiv 1 \pmod{3} \text{ and } \tilde{C} \equiv 3\tilde{B} - 1 \pmod{27}$$

and the case $\tilde{B} \equiv 2 \pmod{3}$ is impossible. Hence, with the conditions of Definition 4.5, we can never be in the case where $(\tilde{B}, \tilde{C}) = (\frac{B}{9}, \frac{C}{27})$ is a pair of integers related to a polynomial in the set $P(X)$. This leads to the statement that minimal pairs (B, C) are characterized by the validity of the inequality $\alpha^2 B(x) \geq B(x)$ which finishes the proof. □

We are now interested in the number of minimal pairs (B, C) or, in other words, in the number of polynomials in the set $P(X)$ which have roots x so that the inequality $B(\alpha x + \beta) \geq B(x)$ is fulfilled for all $\alpha, \beta \in \mathbb{Q}$. The reason for this is that for each pair $(\tilde{B}, \tilde{C})$ which is not minimal, there is a minimal pair (B, C) which defines the same number field (up to isomorphy, the polynomials $f_1(t) = t^3 - 9t + 4$ and $f_2(t) = t^3 - 6t + 3$ from Remark 4.13 illustrate this) while minimal pairs are related uniquely to fields. Speaking in terms of polynomials, this means that we are only considering those polynomials f for which the three distinct real roots are not among the conjugates of the elements of the integral basis of a field defined by another polynomial since such two fields are always isomorphic. The following proposition gives us the asymptotical value of the number of minimal pairs.

Theorem 4.9. *The number of pairs (B, C) fulfilling the above minimality-conditions is asymptotically equal to*

$$\zeta(5)^{-1} \frac{\sqrt{6}}{15} X^{5/2} \tag{4.7}$$

where $\zeta(\cdot)$ is the Riemann zeta function.

Proof. Let $\mathcal{N}(X)$ be the set of all pairs (B, C) where $0 < B \leq \frac{3X}{2}$ and the congruences from above hold. We then know that

$$\mathcal{N}(X) = \bigcup_{k \geq 1} \{(k^2 B, k^3 C) \mid 0 < B \leq \frac{3X}{2}, (B, C) \text{ minimal}\}.$$

Set

$$\mathcal{N}_{\min}(X) := \{(B, C) \mid (B, C) \in \mathcal{N}(X) \text{ and } (B, C) \text{ is minimal}\}.$$

4. Parametrization of the Polynomials

Then $\mathcal{N}(X)$ and $\mathcal{N}_{\min}(X)$ are related via the equalities

$$\mathcal{N}(X) = \sum_{k\geq 1} \mathcal{N}_{\min}\left(\frac{X}{k^2}\right) \tag{4.8}$$

$$\mathcal{N}_{\min}(X) = \sum_{k\geq 1} \mu(k)\mathcal{N}\left(\frac{X}{k^2}\right) \tag{4.9}$$

where $\mu(k)$ is the Möbius function and $\mu(k) = \pm 1$. The sums in (4.8) and (4.9) both converge absolutely since for large enough k the value for both $\mathcal{N}(X/k^2)$ and $\mathcal{N}_{\min}(X/k^2)$ is equal to 0 (because X is fixed).

Using the fact that the number of all pairs (B, C) is asymptotically equal to $\frac{\sqrt{6}}{15}X^{5/2}$ as seen above, we obtain for real constants $c = \frac{\sqrt{6}}{15}X^{5/2}$ and d that

$$\left|\mathcal{N}(X) - cX^{5/2}\right| \leq dX^{3/2}$$

which implies

$$\left|\mathcal{N}_{\min}(X) - c\sum_{k\geq 1} \mu(k)\frac{X^{5/2}}{k^5}\right| \leq d\sum_{k\geq 1} \frac{X^{3/2}}{k^3} = d\zeta(3)X^{3/2}$$

while

$$\sum_{k\geq 1} \mu(k)\frac{X^{5/2}}{k^5} = \zeta(5)^{-1}X^{5/2}$$

where we have used the well-known Dirichlet series

$$\sum_{k\geq 1} \frac{\mu(k)}{k^s} = \zeta(s)^{-1}.$$

This yields (4.7) as the asymptotical number of minimal pairs (B, C). □

Remark 4.10. A more vivid "explanation" of what happens in Theorem 4.9 is the following: recall that the number of all (that means minimal and non-minimal) pairs (B, C) is asymptotically equal to $\frac{\sqrt{6}}{15}X^{5/2}$. Let $p \neq 3$. Then only the first condition in Definition 4.5 is of interest and $p^2 \mid B$ in $\frac{1}{p^2}$ of all choices for B. Similarly, $p^3 \mid C$ in $\frac{1}{p^3}$ of all choices for C. Putting this together

4. Parametrization of the Polynomials

we obtain that $p^2 \mid B$ and $p^3 \mid C$ in $\frac{1}{p^5}$ of all cases. This means that $\frac{1}{p^5}$ of all pairs (B, C) have to be excluded for not being minimal or, equivalently, $1 - \frac{1}{p^5}$ of all pairs will be taken into account for each p. We drop the condition $p \neq 3$ for the moment (we will later see that this is not a problem). Since the first condition then should hold for any prime p, we have to take into account

$$\prod_{p \in \mathbb{P}} \left(1 - \frac{1}{p^5}\right) = \zeta(5)^{-1}$$

of all choices of pairs (B, C) in total. It is left for us to show that for $p = 3$ we have to take into account $1 - \frac{1}{3^5}$ of all pairs (B, C).

Let $p = 3$. Then we exclude $\frac{1}{27} = \frac{1}{3^3}$ of all cases due to $27 \nmid B$ and $\frac{1}{3^6}$ of all cases due to $3^6 \nmid C$. Because of these two condition we therefore exclude

$$\frac{1}{3^3} \cdot \frac{1}{3^6} = \frac{1}{3^9}$$

of all cases. Due to the remaining conditions we furthermore exclude

$$\frac{1}{3^2} \cdot \frac{1}{3^3} \cdot \frac{1}{3} \cdot \frac{2}{27} = \frac{2}{3^9}$$

of all cases. Altogether we therefore exclude $\frac{3}{3^9} = \frac{1}{3^8}$ of all choices for (B, C). Equivalently, $1 - \frac{1}{3^8}$ of all choices have to be taken into account.

Recall the equivalences (4.1) to (4.3). Without 3-minimality we have seen that we have to take $\frac{3}{81} = \frac{1}{27}$ of all choices for (B, C) into account. If one of the two equivalences (4.2) or (4.3) holds (which, as we have seen, is the case in $\frac{2}{81}$ of all cases) then the pair (B, C) is already 3-minimal. On the other hand, if (4.1) holds, we have to exclude $\frac{1}{3^8}$ of the cases. Altogether, requiring 3-minimality, we obtain that we have to consider

$$\frac{2}{81} + \left(\frac{1}{81} - \frac{1}{3^8}\right) = \frac{1}{27} - \frac{1}{3^8}$$
$$= \frac{1}{27}\left(1 - \frac{1}{3^5}\right).$$

4. Parametrization of the Polynomials

However, the factor $\frac{1}{27}$ has already been treated above (see (4.5) and (4.6)). Therefore, we obtain the desired result (4.7).

It takes some effort to decide in general whether two given number fields K_1 and K_2 are isomorphic or not and to compute one or more isomorphisms explicitly although there exist polynomial time algorithms that use polynomial factorization in number fields and the fact that this is possible in polynomial time; the problem can also be decided by using methods of linear algebra. See Cohen, 1993, Chapter 4.5, and Lenstra, 1983, for references. However, in our special situation (due to the restriction $a \in \{0,1,2\}$ and the definition of B and C) we can easily see which of the polynomials in question generate number fields which are isomorphic via an isomorphism of the form $x \mapsto \alpha x + \beta$ and we can as well specify the isomorphism explicitly.

Proposition 4.11. *Polynomials parametrized by (B,C) and $(B,-C)$, respectively, generate isomorphic number fields and the isomorphism is given by $x \mapsto -x$ in the case $a = 0$, and $x \mapsto -x - 1$ in the cases $a = 1$ and $a = 2$.*

Proof. We know that two polynomials f and $-f$ generate the same field in any case because they have the same roots. From this, the assumption is clearly true for $a = 0$ since in that case, (B,C) and $(B,-C)$ define polynomials $f_1(t) = t^3 + bt + c$ and $f_2(t) = t^3 + bt - c$, respectively. If $f_1(t)$ generates the field K then so does $f_2(-t) = -t^3 - bt - c = -f_1(t)$. Strictly speaking, the fields generated by f_1 and f_2, respectively, are isomorphic via the isomorphism $x \mapsto -x$ which makes them technically the same.

Let now $a_1, a_2 \in \{1,2\}$. Let further $f_1(t) = t^3 + a_1 t^2 + b_1 t + c_1$ be the polynomial defined by (B,C) and let $f_2(t) = t^3 + a_2 t^2 + b_2 t + c_2$ be the polynomial defined by $(B,-C)$. Note that necessarily $f_1 = f_2$ if $a_1 = a_2$ since by the definition of B and C we then have $b_1 = b_2$ and $c_1 = c_2$. We therefore set $a_1 = 1$ and $a_2 = 2$ without loss of generality. This implies

$$1 - 3b_1 = 4 - 3b_2$$
$$\Leftrightarrow b_2 = b_1 + 1$$

as well as

$$-2 + 9b_1 - 27c_1 = -(-2 \cdot 2^3 + 9 \cdot 2 \cdot (b_1 + 1) - 27c_2)$$
$$\Leftrightarrow -2 + 9b_1 - 27c_1 = -2 - 18b_1 + 27c_2$$
$$\Leftrightarrow c_2 = b_1 - c_1$$

so that we can write $f_2(t) = t^3 + 2t^2 + (b_1 + 1)t + (b_1 - c_1)$.

If x is a root of $f_1(t)$ then $-x - 1$ is a root of $f_2(t)$ because

$$\begin{aligned} f_2(-t-1) &= (-t-1)^3 + 2(-t-1)^2 + (b_1+1)(-t-1) + (b_1 - c_1) \\ &= -(t^3 + t^2 + bt + c) \\ &= -f_1(t). \end{aligned}$$

With this, if K_1 is the field generated by f_1 and K_2 is the field generated by f_2, the isomorphism between K_1 and K_2 is given by $x \mapsto -x - 1$. This finishes the proof. $\square$

From Theorem 4.9 and Proposition 4.11 we easily derive the following result.

Corollary 4.12. *The number of isomorphism classes of number fields generated by polynomials which are defined by pairs (B, C) that fulfill the above conditions is asymptotically equal to*

$$\frac{1}{2}\zeta(5)^{-1}\frac{\sqrt{6}}{15}X^{5/2}. \tag{4.10}$$

Proof. From the bounds for C given in (4.4) we immediately see that for any choice of B there are the same number of choices for C and $-C$, respectively. The claim follows from combining (4.7) with the result of Proposition 4.11. $\square$

Remark 4.13. Note that isomorphisms of another form than $x \mapsto \alpha x + \beta$ are not "covered" by the reduction to minimal pairs (B, C). For $X = 18$ we obtain a list which contains the polynomials $f_1(t) = t^3 - 9t + 4$ and $f_2(t) = t^3 - 6t + 3$ while the number fields K_1 generated by f_1 and K_2 generated by f_2 are isomorphic via the mapping $x \mapsto x^2 + x - 4$.

4. Parametrization of the Polynomials

Definition 4.14 (Correction factor). We define the *correction factor* as

$$\begin{aligned}
\kappa &:= \frac{1}{2}\zeta(5)^{-1}\frac{\sqrt{6}}{15} \\
&= \frac{\sqrt{6}}{30}\zeta(5)^{-1} \\
&\approx 0.0787418966.
\end{aligned}$$

Definition 4.15. We define $P_\kappa(X)$ as the set which includes the $\kappa X^{5/2} + \mathcal{O}(X^{3/2})$ pairs (B, C) that are related to the distinct isomorphism classes of number fields as described above.

Remark 4.16. Note that we could also interpret $P_\kappa(X)$ as the set of $\kappa X^{5/2} + \mathcal{O}(X^{3/2})$ polynomials which generate distinct isomorphism classes of number fields.

We will later see (computationally) that the number of isomorphism classes of fields generated by polynomials of the set $P(X)$ is asymptotically equal to $\kappa X^{5/2}$ and we will evaluate how fast the convergence is.

4.2. A Bound for the Number of Galois Fields

Before we come to the computational part of the theory which will give an idea of how fast the convergence of the fraction $|P_\kappa(X)|/|F(X)|$ is, we have to prove that the fraction actually converges (with limit 1). For this purpose we have to take a closer look at a subset of all fields which we are considering, namely the set of all of these fields which are Galois. We will see that Galois number fields are counted three times as often in the above counting procedure for pairs (B, C) as non-Galois fields. Together with the important result that the number of Galois number fields in question can be neglected since it is only of an order smaller than $\mathcal{O}(X^{3/2})$ and with the above computations, this will give the desired result. We start with the necessary definitions of separable and normal field extensions and with a lemma which is useful in the special case of totally real cubic fields.

Definition 4.17 (Separable field extension). A field extension $K \supseteq \mathbb{Q}$ is called a *separable extension* of $\mathbb{Q}$ if and only if for every $\theta \in K$ the minimal polynomial of θ over $\mathbb{Q}$ has three distinct roots.

Definition 4.18 (Normal field extension). A field extension $K \supseteq \mathbb{Q}$ is called a *normal extension* of $\mathbb{Q}$ if and only if the minimal polynomial over $\mathbb{Q}$ of every element in K splits over K.

Definition 4.19 (Galois extension). A field extension $K \supseteq \mathbb{Q}$ is called a *Galois extension* if and only if it is a separable extension and a normal extension.

Lemma 4.20. *For an algebraic number field K of degree 3 the following statements are equivalent:*

1. *K is Galois,*

2. *K is cyclic,*

3. *K is abelian.*

Proof. This is clear due to the fact that the Galois group of K must be isomorphic to $\mathbb{Z}/3\mathbb{Z}$. □

The next lemma, which states that in the following we only have to deal with the question whether a given K is normal or not, and its proof are based on Bhattacharya et al., 1994, Corollary 3.5.

Lemma 4.21. *Every algebraic number field $K \supseteq \mathbb{Q}$ is separable.*

Proof. Let $f(t) = \sum_{i=0}^{n} a_i x^i$ be an irreducible polynomial over $\mathbb{Q}$. We know that f has multiple roots if and only if $f'(x) = \sum_{i=1}^{n} i a_i x^{i-1} = 0$. This implies $i a_i = 0$ for all $i \in \{1, \ldots, n\}$. Since $\text{char}(\mathbb{Q}) = 0$ this means that $a_i = 0$ for all $i \in \{1, \ldots, n\}$. It follows that $f = a_0 \in \mathbb{Q}$ which is a contradiction. Hence, all roots of f are simple. Since this holds in particular for any minimal polynomial of an element $\theta \in K$, we obtain the desired result. □

Lemma 4.22. *Let f be a defining polynomial for a cubic number field $K \supseteq \mathbb{Q}$. Then either all three or only one of the roots of f lie in K.*

4. Parametrization of the Polynomials

Proof. Recall that f is a defining polynomial for K if $K = \mathbb{Q}(x)$ for a root x of f. If K is a Galois extension of $\mathbb{Q}$ then K is separable and normal. This is equivalent to the statement that the minimal polynomial over $\mathbb{Q}$ of every element $\theta \in K$ splits over K. In that case, we can write f as the product of three polynomials of degree 1 which obviously all have one root in K.

If K is non-Galois then, as a finite field extension, it is still separable but not normal. Hence, there are elements in K whose minimal polynomial over $\mathbb{Q}$ does not split over K. Assume there are two roots of f which lie in K and the third one does not. Then $f = gh$ for a polynomial g of degree 1 and a polynomial h of degree 2 with h being the polynomial with roots that lie in K. Hence, for degree reasons, the root of f which does not lie in K is the root of g which is a polynomial of degree 1. This is a contradiction and therefore either all three or only one of the roots of f lie in K, depending on whether K is Galois or not. □

Remark 4.23. Note that the only possibilities for Galois groups of separable and irreducible cubic polynomials f are A_3 and S_3. Therefore, Lemma 4.22 means that if the Galois group of a cubic polynomial f over $\mathbb{Q}$ is A_3 (K is Galois), all roots of f generate the same field.

Lemma 4.22 tells us what the difference between Galois fields and fields that are not Galois is with respect to the set of pairs (B, C). This statement is formalized in the following proposition.

Proposition 4.24. *A pair (B, C) which defines a Galois number field K is counted three times as often in the above counting procedure as a pair (B, C) which defines a non-Galois number field K.*

Proof. This is an easy consequence of Lemma 4.22. If K is Galois then the respective pair $(B, C) = (B(x), C(x))$ is counted three times in the above procedure—once for every root x of f that lies in K because the values of B and C are independent of x. Similarly a pair (B, C) is counted only once for non-Galois K since then only one root x of f lies in K. □

Remark 4.25. Note that there can still be many pairs (B,C) related to the same field K. We will investigate this fact (and see that it is nothing we have to worry about) in Chapter 4.4.

Consider now pairs (K,x) where $x \in \mathcal{O}_K/\mathbb{Z}, x \notin \mathbb{Z}$. We say that $(K,x) \sim (K',x')$, in words "the pairs are equivalent", if there is a $\sigma : F \xrightarrow{\sim} F'$ with $\sigma(x) = x'$. We conclude from this, the above considerations, and Lemma 4.22 that the correspondence $(K,x) \mapsto (B,C)$ is injective $(1-1)$ for non-Galois K and $3-1$ for Galois K. It is now left to show that the number of considered K which are Galois is of an order less than $\mathcal{O}(X^{5/2})$ and that the number of pairs (K,x) is of negligible order for a fixed K (we will see this in Chapter 4.4) because then we will obtain the desired result

$$\lim_{X \to \infty} \frac{|P_\kappa(X)|}{|F(X)|} = 1.$$

Indeed, the number of Galois fields will turn out to be of the order $\mathcal{O}(X^{3/2})$ due to an upper bound for the field discriminant of K. This will be shown in a rigorous way now and we start with a lemma as preparation.

Lemma 4.26. *For all $x, y, z \in \mathbb{R}$ the inequality*

$$(x-y)^2(x-z)^2(y-z)^2 \leq \frac{1}{2}\left(x^2 + y^2 + z^2 - \frac{1}{3}(x+y+z)^2\right)^3 \quad (4.11)$$

holds.

Proof. Consider the vector $(x,y,z) \in \mathbb{R}^3$. First, note that both sides of the inequality are invariant under translation by a constant, that means by vectors $(v,v,v) \in \mathbb{R}^3$. For the left-hand side this is obvious because

$$\begin{aligned}
& (x+v-y-v)^2(x+v-z-v)^2(y+v-z-v)^2 \\
= & (x-y+v-v)^2(x-z+v-v)^2(y-z+v-v)^2 \\
= & (x-y)^2(x-z)^2(y-z)^2
\end{aligned}$$

4. Parametrization of the Polynomials

while for the right-hand side we have

$$x^2 + y^2 + z^2 - \frac{1}{3}(x+y+z)^2$$
$$= \frac{2}{3}\left(x^2 - xy - xz + y^2 - yz + z^2\right)$$

as well as

$$(x+v)^2 + (y+v)^2 + (z+v)^2 - \frac{1}{3}(x+v+y+v+z+v)^2$$
$$= x^2 + y^2 + z^2 + 2xv + 2yv + 2zv + v^2 + v^2 + v^2$$
$$-3v^2 - 2vx - 2vy - 2vz - \frac{1}{3}x^2 - \frac{1}{3}y^2 - \frac{1}{3}z^2 - \frac{2}{3}xy - \frac{2}{3}xz - \frac{2}{3}yz$$
$$= \frac{2}{3}\left(x^2 - xy - xz + y^2 - yz + z^2\right).$$

With this, we can choose $(v, v, v) = -(z, z, z)$ and therefore set $z = 0$ to be able to rewrite the inequality as

$$\frac{1}{2}\left(x^2 + y^2 - \frac{1}{3}(x+y)^2\right)^3 - (x-y)^2 x^2 y^2 \geq 0.$$

By the method of Lagrange multiplicators, computing the extremes of the second addend

$$(x-y)^2 x^2 y^2$$

upon condition that the first addend

$$\frac{1}{2}\left(x^2 + y^2 - \frac{1}{3}(x+y)^2\right)^3$$

is constant, which yields $x + y = 0$, we obtain that

$$\frac{1}{2}\left(x^2 + y^2 - \frac{1}{3}(x+y)^2\right)^3 - (x-y)^2 x^2 y^2 < 0 \qquad (4.12)$$

has no solution. Therefore (4.11) holds for all $x, y, z \in \mathbb{R}$. $\square$

We are now able to give an upper bound for the discriminant of a cubic field in terms of its second successive minimum (and therefore in terms of X) which will be needed in further considerations.

4. Parametrization of the Polynomials

Proposition 4.27. *For a cubic field K with discriminant $d(K)$ and second successive minimum $M_2(K) \leq X$ we have the inequality*

$$d(K) \leq \frac{1}{2}M_2(K)^3 \leq \frac{1}{2}X^3.$$

Proof. As seen above, the inequality $M_2(K) \leq \operatorname{Tr}(x^2)$ always holds. Together with the requirement $\operatorname{Tr}(x^2) \leq X$, this gives the second inequality. For the first inequality note that a totally real cubic field K is generated by an element $x \in \mathcal{O}_K \setminus \mathbb{Z}$ or, in other words, the minimal polynomial f of x generates $K = \mathbb{Q}(x)$, that is, f is a defining polynomial for K. Therefore, for some $c \in \mathbb{N}$, we have

$$d(K) = \frac{d(f)}{c^2} \leq d(f) = \prod_{1 \leq i < j \leq 3}(x_i - x_j)^2 \leq \frac{1}{2}\left(\operatorname{Tr}(x^2) - \frac{1}{3}\operatorname{Tr}(x)^2\right)^3$$

by applying Lemma 4.26 to the conjugates x_1, x_2, x_3 of x (which are the three distinct real roots of f). This directly leads to the desired inequality

$$d(K) \leq \frac{1}{2}\left(\operatorname{Tr}(x^2) - \frac{1}{3}\operatorname{Tr}(x)^2\right)^3 \leq \frac{1}{2}M_2(K)^3$$

which finishes the proof. $\square$

In Cohn, 1954, we see that the number of abelian (which implies Galois in our special case) fields with discriminant less than or equal to D, denoted $N_A(D)$ is of the order $\mathcal{O}(D^{1/2})$. More precisely we have the following proposition.

Proposition 4.28. *For the number $N_A(D)$ of abelian cubic fields K with $d(K) \leq D$ we have*

$$\lim_{D \to \infty} N_A(D) = \frac{11\gamma}{18}D^{1/2}$$

where $\gamma \approx 0.28$.

The precise value for γ is given by Cohn but not of further interest for our purposes. All we have to memorize is that $N_A(D) = \mathcal{O}(D^{1/2})$. With this, we are now able to compute the number of Galois fields in question.

4. Parametrization of the Polynomials

Corollary 4.29. *The number of polynomials defining Galois fields among the set $P_\kappa(X)$ is of the order $\mathcal{O}(X^{3/2})$.*

Proof. Combining Proposition 4.28 and Proposition 4.27, we obtain

$$N_A(D) = \mathcal{O}(D^{1/2}) \leq \mathcal{O}((X^3)^{1/2}) = \mathcal{O}(X^{3/2})$$

where we have omitted the factor 1/2 from above (which would have been eliminated by the $\mathcal{O}$ anyway). Writing the second inequality as an equality, which is okay in the $\mathcal{O}$-notation, we obtain the desired result. □

We can therefore neglect all fields among those in question which are Galois since there are only $\mathcal{O}(X^{3/2})$ many of them which is less than $\mathcal{O}(X^{5/2})$. An interesting result which provides a trivial algorithm to check whether a given cubic field K is Galois or not is given in Cohen, 1993, Proposition 6.4.3. We will present the proposition with a rough sketch of the proof.

Proposition 4.30. *Let $K = \mathbb{Q}(x)$ be a cubic field where x is an algebraic integer with minimal polynomial $P(X)$. Then K is Galois over $\mathbb{Q}$ if and only if the discriminant of $P(X)$ is a square.*

Proof. The only transitive subgroups of S_3 are $C_3 \simeq A_3$ and $S_3 \simeq D_3$. If f is the defining polynomial for the field K then we have either $\mathrm{Gal}(f) \simeq C_3$ or $\mathrm{Gal}(f) \simeq S_3$ with the first isomorphy holding if $d(f) = e^2$ for some e and the second one holding if $d(f)$ is not a square. The result follows immediately. □

4.3. An Alternative Approach to the Bound for the Number of Galois Fields

The fact that a cubic field is Galois if and only if the discriminant of its defining polynomial is a square as proved in Proposition 4.30 suggests another approach to the task of determining the number of Galois fields as described above.

An equality for the number $n_3(D)$ of cubic extensions of $\mathbb{Q}$ with fixed discriminant D has been introduced by Hasse, 1930. An upper bound for this number, namely $n_3(|D|) = \mathcal{O}(X^{0.44178\cdots})$ has been given by Helfgott and Venkatesh, 2006, and has been improved to $n_3(|D|) = \mathcal{O}(X^{1/3+\varepsilon})$ by Ellenberg and Venkatesh, 2007. In Wong, 1999, it is shown that, assuming both the Birch-Swinnerton-Dyer conjecture and the generalized Riemann hypothesis, $h_3(D) = \mathcal{O}(|D|^{1/4+\varepsilon})$ where $h_3(D)$ is the number of elements in the class group of a quadratic field $\mathbb{Q}(\sqrt{D})$ whose cube is the principal ideal class. This implies $n_3(D) = \mathcal{O}(|D|^{1/4+\varepsilon})$. Wong's result allows us to give an asymptotic value for the number of Galois fields among the set $P_\kappa(X)$, namely $\mathcal{O}(X^{9/4+\varepsilon})$.

First, we present the result of Wong as a proposition.

Proposition 4.31. *Assuming the Birch-Swinnerton-Dyer conjecture and the generalized Riemann hypothesis, the number $n_3(D)$ of cubic extensions of $\mathbb{Q}$ with fixed dicriminant D is given by*

$$n_3(D) = \mathcal{O}\left(|D|^{1/4+\varepsilon}\right).$$

With the following corollary we get another proof of the fact that the number of Galois fields in $P_\kappa(X)$ can be neglected. However, this method has to be treated carefully since it depends on the assumption that both the Birch-Swinnerton-Dyer conjecture and the generalized Riemann hypothesis are true.

Corollary 4.32. *The number of fields K with $M_2(K) \leq X$ and $d(K) = e^2$ for some integer e is of the order $\mathcal{O}(X^{3/2})$.*

Proof. This is clear since $\sqrt{X}$ is an upper bound for the number of squares less than or equal to X. □

The next corollary is now a trivial consequence of Corollary 4.32 and Propositions 4.30 and 4.31.

Corollary 4.33. *Assuming the Birch-Swinnerton-Dyer conjecture and the generalized Riemann hypothesis, the number of Galois number fields among the set $P_\kappa(X)$ is of the order $\mathcal{O}(X^{9/4+\varepsilon})$.*

4. Parametrization of the Polynomials

Proof. For a fixed X there are $\mathcal{O}(X^{3/2})$ distinct discriminants which belong to Galois fields and $\mathcal{O}(X^{3/4+\varepsilon})$ different fields with the same discriminant so that there are

$$\mathcal{O}(X^{3/2}) \cdot \mathcal{O}(X^{3/4+\varepsilon}) = \mathcal{O}(X^{9/4+\varepsilon})$$

Galois fields among the set $P_\kappa(X)$ in total. $\square$

Remark 4.34. Note that once a bound $n_3(D) = \mathcal{O}(|D|^k)$ with $k < 1/3$ can be found (for the special case of totally real cubic fields) without assuming unproved statements, the above argumentation works with this bound and provides a rigorous proof that the number of Galois fields among the set $P_\kappa(X)$ is negligible.

Most of the bounds cited here as well as the 3-part of the class number of quadratic fields $h_3(D)$ and its consequences for our special case of cubic fields are treated in detail in Pierce, 2005. It is conjectured by Pierce (and other authors) that $h_3(D) \ll |D|^\varepsilon$ for any $\varepsilon > 0$. Note however that it is everything but trivial to obtain a better bound than $h_3(D) = |D|^{1/2+\varepsilon}$ and by today's knowledge an unconditional bound with $k < 1/3$ in Remark 4.34 is not within touching distance.

4.4. Fields with more than one related Minimal Pair (B, C)

With the above considerations and some more calculations we can now see more clearly that the number of minimal pairs (B, C) is exactly what we are interested in. This is because the number of fields with more than one related (minimal) pair (B, C) is of negligible order $\mathcal{O}(X^2)$. We start with the definition of a primitive vector and, based on that, we bound the number of fields with the above property in terms of the discriminant.

Definition 4.35 (Primitive vector). Let L be a free $\mathbb{Z}$-module isomorphic to $\mathbb{Z}^n$. A vector $v \in L$ is called *primitive* if and only if

$$v \notin \bigcup_{\mathbb{N} \ni m \geq 2} mL.$$

Note the following facts which are easy to see.

Lemma 4.36. *Let $K = \mathbb{Q}(x)$ for a root x of a monic irreducible polynomial f. The positive definite quadratic form $Q(x) = \mathrm{Tr}(x^2) - \frac{1}{3}\mathrm{Tr}(x)^2 = \frac{2}{3}B(x)$ on $\mathcal{O}_K$ has discriminant $d(Q) = \frac{1}{3} \cdot d(K)$ and one of the following three cases holds.*

1. *There is no $x \in \mathcal{O}_K/\mathbb{Z}, x \notin \mathbb{Z}$ with $Q(x) \leq X$,*

2. *There is one pair $\pm x_0$ of (non-trivial) primitive vectors with this property,*

3. *There are two linearly independent (non-trivial) primitive vectors with this property.*

The interesting case in Lemma 4.36 is of course the third one since this case is the only one where it is possible that there are two or more (minimal) pairs (B, C) for one field K. If we consider a reduced basis $(1, \omega_1, \omega_2)$ then $\omega_2 > X$ implies that all $x \in \mathcal{O}_K/Z$ with $Q(x) \leq X$ are multiples of some x_0 with $Q(x_0) = \omega_1$. But then again, this implies that in the case where we can have more than one minimal pair (B, C) for one field K, we have $d(K) \leq 3X^2$ as an upper bound for the field discriminant.

The following lemma gives an upper bound for the number of pairs related to a fixed field K if we are in case 3 in Lemma 4.36. Its proof uses geometric arguments to bound the number of lattice points $v \in \mathbb{Z}^2$ within an ellipse. More information about the (more general) topic of conic sections, particularly about the notions of the discriminant and area of an ellipse, can be found in Salmon, 1960.

4. Parametrization of the Polynomials

Lemma 4.37. *For a positive definite binary quadratic form $Q(x,y) = ax^2 + bxy + cy^2$ we have*

$$|\{v \in \mathbb{Z}^2 \mid Q(v) \leq X\}| \leq \frac{\pi X}{d(Q)^{1/2}} \left(1 + \left(\frac{Y}{X}\right)^{1/2}\right)^2 \quad (4.13)$$

where

$$Y = \max_{|x|,|y| \leq \frac{1}{2}} Q(x,y) = \frac{1}{4}(a + c + |b|).$$

Proof. First, define the discriminant of Q as

$$d(Q) := ac - \frac{b^2}{4}.$$

Note that $d(Q)$ is usually defined as $4ac - b^2$. We multiply this by $\frac{1}{4}$ in order to get rid of the factor 2 in the numerator in the common definition of the area of an ellipse $ax^2 + bxy + cy^2 = 1$ which is given by $A = \frac{2\pi}{\sqrt{4ac-b^2}}$.

Consider now squares $v + [-\frac{1}{2}, \frac{1}{2}] \times [-\frac{1}{2}, \frac{1}{2}]$ for each lattice point $v \in \mathbb{Z}^2$ with $Q(v) \leq X$. Then, if A_v is the area of the square around v, it follows for the sum of the areas of all such squares that

$$\sum_{\substack{v \in \mathbb{Z}^2 \\ Q(v) \leq X}} A_v = \sum_{\substack{v \in \mathbb{Z}^2 \\ Q(v) \leq X}} 1 = |\{v \in \mathbb{Z}^2 \mid Q(v) \leq X\}|$$

which is the left-hand side in (4.13). To obtain the desired inequality, we consider vectors $\tilde{v} = v + w$ with $w \in [-\frac{1}{2}, \frac{1}{2}] \times [-\frac{1}{2}, \frac{1}{2}]$ and with the triangle inequality we obtain

$$Q(\tilde{v})^{1/2} \leq Q(v)^{1/2} + Q(w)^{1/2} \leq X^{1/2} + Y^{1/2}$$

which leads to

$$Q(\tilde{v}) \leq \left(X^{1/2} + Y^{1/2}\right)^2 = X\left(1 + \left(\frac{Y}{X}\right)^{1/2}\right)^2.$$

Using the fact that (with the above definition of $d(Q)$) the area of the ellipse defined by $Q(v) \leq X$ is equal to $\frac{\pi X}{d(Q)^{1/2}}$ and comparing the areas computed above, this yields the desired inequality (4.13). □

We make use of Lemma 4.37 as follows: if Q is reduced and $a, c \leq X$ then $|b| \leq a \leq X$ which implies $Y \leq \frac{3}{4}X$ so that the left-hand side in (4.13) is at most equal to

$$\frac{\pi X}{d(Q)^{1/2}} \left(1 + \frac{\sqrt{3}}{2}\right)^2.$$

Since $Q(x) = \frac{2}{3}B(x)$, this will give the desired bound for the number of pairs related to a fixed field K.

The following result, which is mainly due to Davenport and Heilbronn, 1969, Davenport and Heilbronn, 1971, and has been improved (which is not relevant for our applications) by Shintani, 1975, finally allows us to see that the number of minimal pairs (B, C) is in fact asymptotically equal to the number of fields in question. We will not give the quite lengthy proof for this important result here but refer to Cohen, 2000, Chapter 8.5, where this problem is investigated in detail for the case of totally real cubic fields (which is the one we are interested in).

Proposition 4.38. *Let $N(X)$ be the number of cubic fields K with discriminant $d(K) \leq |X|$. Then $N(X) = \mathcal{O}(X)$.*

Remark 4.39. Better estimates than $\mathcal{O}(X)$ have been given by Davenport and Shintani and the actual value of $N(X)$ (more precisely the value of the remainder term of the number $N_3^+(X)$ of isomorphism classes of totally real cubic fields, see again Cohen, 2000, Chapter 8.5 for reference) has also been treated in Belabas, 1999.

Putting the results of Lemma 4.37 and Proposition 4.38 together and using the following lemma, we obtain the desired result in Proposition 4.41.

4. Parametrization of the Polynomials

Lemma 4.40. *Let $N_3^+(X)$ be the number of isomorphism classes of totally real cubic fields K with $d(K) \leq X$. Then $N_3^+(X) = \mathcal{O}(X)$ and*

$$\sum_{K:d(K)\leq X} \frac{1}{d(K)^{1/2}} = \mathcal{O}(X^{1/2}). \tag{4.14}$$

Proof. The equality $N_3^+(X) = \mathcal{O}(X)$ follows from Proposition 4.38 (see Cohen, 2000, Theorem 8.5.6 for more details and a more precise bound). For the proof of equality (4.14), note that

$$N_3^+(X) = \sum_{k=1}^{X} n_3^+(k)$$

where $n_3^+(k)$ is the number of isomorphism classes of totally real cubic fields K with $d(K) = k$. We then have

$$\sum_{k=1}^{X} \frac{n_3^+(k)}{k^{1/2}} = \frac{N_3^+(X)}{X^{1/2}} + \sum_{k=1}^{X-1} \left(\frac{N_3^+(k)}{(k+1)^{1/2}} \left(\left(1 + \frac{1}{k}\right)^{1/2} - 1 \right) \right)$$

where the first term on the right-hand side is $\mathcal{O}(X^{1/2})$ and each addend is of the order $\mathcal{O}(k^{-1/2})$ which implies equality (4.14). □

Proposition 4.41. *The number of fields with more than one related minimal pair (B, C) is $\mathcal{O}(X^2)$ and can therefore be neglected.*

Proof. This is a consequence of Lemma 4.37 and Proposition 4.38 since they imply that for a real constant c,

$$\sum_{K:d(K)\leq 3X^2} \frac{cX}{d(K)^{1/2}} = \mathcal{O}(X^2)$$

which is due to Lemma 4.40. The facts that $Q(x) = \frac{2}{3}B(x)$ and that $d(K)$ and $d(Q)$ are of the same order (in X) finally give the desired result. □

Remark 4.42. Note that the less sharp bound for $d(K)$ from above, $d(K) = \mathcal{O}(X^3)$, does not work here, because it would yield $\mathcal{O}(X^{5/2})$ as the number of fields with more than one related pair (B, C) instead of $\mathcal{O}(X^2)$. While the influence of a subset of size $\mathcal{O}(X^2)$ is asymptotically 0 for a set of size $\mathcal{O}(X^{5/2})$, this is clearly false if the subset itself already has size $\mathcal{O}(X^{5/2})$.

Now, after finishing the discussion of the theory and showing the desired asymptotic behaviour of the set of minimal pairs $P_\kappa(X)$, we come to the computational part of this thesis where we investigate how fast the convergence of the fraction

$$\lim_{X \to \infty} \frac{|P_\kappa(X)|}{|F(X)|} = 1$$

actually is and which effect the considered errors have, particularly for small values of X.

5. The Rate of Convergence

The conjecture that has been proven theoretically above is that for $X \to \infty$ there are $\kappa X^{5/2}$ isomorphism classes of number fields K with $M_2(K) \leq X$, that is

$$\lim_{X \to \infty} \frac{|P_\kappa(X)|}{|F(X)|} = \lim_{X \to \infty} \frac{\kappa X^{5/2}}{|F(X)|} = 1.$$

It is interesting to investigate how fast the convergence of this fraction is and we will investigate this rate of convergence computationally by checking the results for different values of X. We compute the number of fields with $M_2(K) \leq X$ by using a quite simple algorithm which we present now. To make reading easier, we write

$$\begin{aligned}
\underline{b} &:= \lceil \frac{a^2 - X}{2} \rceil, \\
\overline{b} &:= \lfloor \frac{a^2}{3} \rfloor, \\
\underline{c} &:= \lceil -\frac{1}{27} \left(2(a^2 - 3b)^{3/2} + (2a^3 - 9ab) \right) \rceil, \\
\overline{c} &:= \lfloor \frac{1}{27} \left(2(a^2 - 3b)^{3/2} - (2a^3 - 9ab) \right) \rfloor
\end{aligned}$$

for the lower and upper bounds for the polynomial coefficients (defined by X) as computed above.

5.1. The Main Algorithm

Algorithm 1 Compute the number of fields K with $M_2(K) \leq X$
---
input: X
  $L \leftarrow \{\}$ (an empty list)
  for $a = 0$ to 2 **do**
    for $b = \underline{b}$ to $\overline{b}$ **do**
      for $c = \underline{c}$ to $\overline{c}$ **do**
        $f(t) \leftarrow t^3 + at^2 + bt + c$
        if f is irreducible and the field generated by f is not isomorphic to any field generated by a polynomial in L **then**
          add f to L
        end if
      end for
    end for
  end for
  return $|L|$

The algorithm has been implemented in PARI (see appendix for source codes) and it has been tested, that means its output has been verified (for small X) by computing $M_2(K)$ for all K in question from the QaoS database (see Freundt, 2005). We have used the estimate $d(K) \leq \frac{1}{2}M_2(K)^3$ from Proposition 4.27[5] to limit the number of results of the database query since the second successive minima of the fields are not contained in the database while the discriminants of course are. Table 4 shows an overview of the computational results. It contains the values of X and the resulting numbers of polynomials, non-isomorphic fields, the value $\kappa X^{5/2}$, the value of the fraction $\kappa X^{5/2}/\#$Fields as well as the value ρ_X which is defined as follows.

---
[5] This estimate is actually a very sharp one, the value of the difference between $\frac{1}{2}X^3$ and the largest field discriminant among the set of fields defined by X is almost always only 27 and for all values of $X \leq 1000$ we applied Algorithm 2 to, it is never larger than 675.

5. The Rate of Convergence

For a number field K set $\rho_K := 2^2 h(K) R(K)/d(K)^{1/2}$ where $h(K)$ is the class number, $R(K)$ is the regulator and $d(K)$ is, as before, the discriminant of K. We then set

$$\rho_X := \sum_K \rho_K$$

where the sum runs over all non-isomorphic K that are generated by the set of polynomials $P_\kappa(X)$ defined by X. To compute the values of ρ_X in addition to the number of fields, we have to extend Algorithm 1 slightly to

---
Algorithm 2 Compute the number of fields K with $M_2(K) \leq X$ together with ρ_X
---
input: X
  $L \leftarrow \{\}$ (an empty list)
  $\rho_X \leftarrow 0$
  for $a = 0$ to 2 **do**
    for $b = \underline{b}$ to $\overline{b}$ **do**
      for $c = \underline{c}$ to $\overline{c}$ **do**
        $f(t) \leftarrow t^3 + at^2 + bt + c$
        if f is irreducible and the field generated by f is not isomorphic to
        any field generated by a polynomial in L **then**
          add f to L
          compute the field K defined by f
          $h(K) \leftarrow$ the class number of K
          $R(K) \leftarrow$ the regulator of K
          $d(K) \leftarrow$ the discriminant of K
          $\rho_K \leftarrow 2^2 h(K) R(K)/d(K)^{1/2}$
          $\rho_X \leftarrow \rho_X + \rho_K$
        end if
      end for
    end for
  end for
  return $|L|$ and ρ_X

The values for ρ_K are related to the class number formula, which is presented in the following section.

5.2. The Class Number Formula and Other Convergences

We will now introduce the class number formula which presents a connection between several invariant values of a number field, namely its class number, regulator, discriminant, and signature and the Dedekind zeta function. The formula is presented without a proof of the limit and we will only elaborate shortly on its connection to our theory. We use the concept to present another fraction of numbers which appear in the course of computational investigations and which seems to converge a bit faster than the original fraction we are interested in. Furthermore, another fraction which is not directly related to the values ρ_K or ρ_X but whose convergence seems also interesting is introduced.

Theorem 5.1. *For the class number h_K of K we have*

$$\lim_{s\to 1}(s-1)\zeta_K(s) = \frac{2^{r_1} \cdot (2\pi)^{r_2} \cdot h_K \cdot R_K}{w_K \cdot d(K)^{1/2}} \tag{5.1}$$

where r_1 is the number of real embeddings of K, r_2 is the number of complex embeddings of K, w_K is the number of roots of unity in K and h_K, R_K and $d(K)$ are as above the class number, regulator and discriminant of K, respectively. $\zeta_K(s)$ is the Dedekind zeta function of K.

Remark 5.2. Since we have $r_1 = 3$, $r_2 = 0$ and $w_K = 2$, equality (5.1) reads

$$\lim_{s\to 1}(s-1)\zeta_K(s) = \frac{2^2 \cdot h_K \cdot R_K}{d(K)^{1/2}}$$

in our special case so that

$$\rho_K = \lim_{s\to 1}(s-1)\zeta_K(s)$$

and

$$\rho_X = \lim_{s\to 1}(s-1)\sum_K \zeta_K(s).$$

5. The Rate of Convergence

With Remark 5.2, we have (on the verge of our theory) introduced a relation between the Dedekind zeta function and the number of isomorphism classes of fields subject to the choice of X. This relation will however not be investigated in more detail although an interesting convergence can be observed. Taking into account the value of the fraction

$$\frac{\rho_X}{X^{5/2}}$$

we obtain a slightly faster convergence than we have for the original fraction. Figure 2 and Table 2 illustrate this.

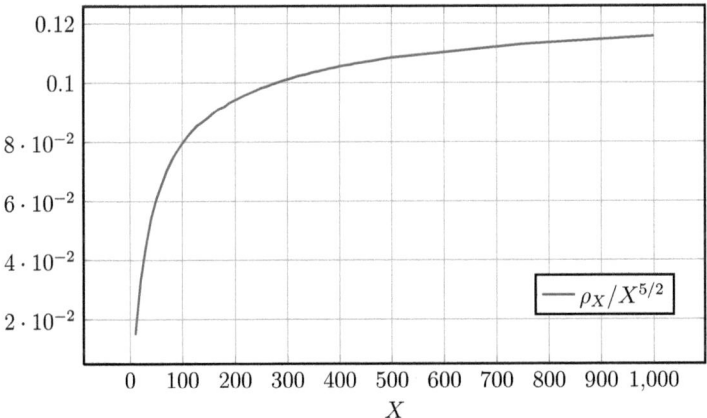

Figure 2.: Value of the fraction $\frac{\rho_X}{X^{5/2}}$

5. The Rate of Convergence

X	Value
10	0.015
100	0.079
200	0.094
300	0.101
400	0.106
500	0.108
750	0.113
1000	0.116

Table 2.: Value of the fraction $\frac{\rho_X}{X^{5/2}}$

Another fraction, which is not related to ρ_X directly, but obviously also converges pretty fast in comparison to the original fraction, is

$$\frac{\kappa X^{5/2} - |F(X)|}{X^2}$$

with X^2 being the order of the number of fields with more than one related pair (B, C) as seen in Chapter 4.4. The convergence of this fraction for $X \to \infty$ is illustrated in Figure 3[6] and Table 3.

---

[6] The values in Figure 3 have been scaled by the factor 10^2 for better readability of the axis, note also that there are no known values of the fraction visualized in Figure 3 for $500 < X < 750$ and $750 < X < 1000$ which explains the jumps in the graph for $X \leq 500$ which are missing for $X > 500$.

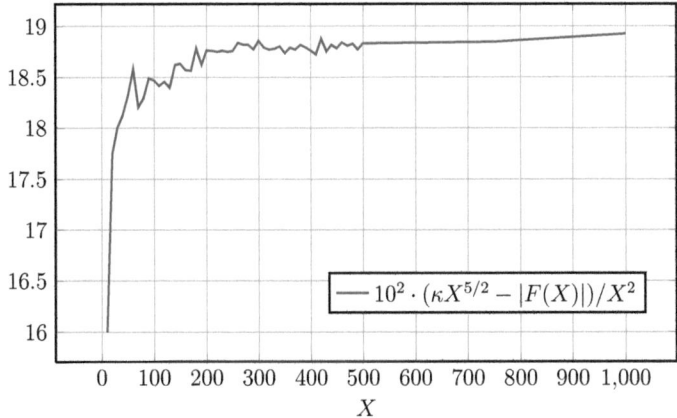

Figure 3.: Value of the fraction $10^2 \cdot \frac{\kappa X^{5/2} - |F(X)|}{X^2}$

X	Value
10	0.1600
100	0.1847
200	0.1876
300	0.1886
400	0.1876
500	0.1883
750	0.1885
1000	0.1893

Table 3.: Value of the fraction $\frac{\kappa X^{5/2} - |F(X)|}{X^2}$

5.3. Technical Details and an Overview of the Computational Results

The PARI-implementation of Algorithm 2 can also be found in the appendix. The problem with this naive implementations of the algorithms is that they are very slow due to a bottleneck in the isomorphism-check for two given number fields. For an input value of $X = 500$ the running time is even in the order of a few days on an average modern personal computer although simple optimizations (that do not show in the pseudocode representations above) have already been treated in the PARI code. A faster implementation of the algorithms does not yet exist but would immediately exist, as stated before, once a faster isomorphism-check could be found and implemented either in general or for the special case of totally real cubic fields. However, the results for some values of X are known and at least they point to the right direction. In Table 4, the values for $\kappa X^{5/2}$ are rounded to the nearest integer.

X	#Polynomials	#Fields	$\kappa X^{5/2}$	Ratio	ρ_X
10	62	9	25	2.778	4.737
50	2930	934	1392	1.490	1069.057
100	16415	6027	7874	1.306	7941.241
150	45136	17506	21699	1.240	24361.036
200	92569	37038	44543	1.203	53288.154
250	161632	66091	77814	1.177	97061.572
300	254883	105772	122746	1.160	157612.104
350	374634	157505	180458	1.146	237507.556
400	523025	221952	251974	1.135	337806.188
500	913503	393105	440181	1.120	606152.540
750	2516676	1106971	1212995	1.096	1741855.204
1000	5165625	2300779	2490037	1.082	3661005.359

Table 4.: Overview: Computational results

A more detailed version of Table 4 is given in the appendix. It contains some more lines as well as some additional information in each line. All computations have been carried out using GP/PARI CALCULATOR Version 2.3.4 (32 bit) on a 64 bit Windows 2008 Server employing AMD Opteron 2350 processors at 2.00 GHz. Due to the implementation of GP/PARI CALCULATOR only one processor kernel could be used at a time. Using the PARI library within a sophisticated program written in a very fast programming language such

5. The Rate of Convergence

as C or C++ could significantly improve the running time of the program. This, however, has not been checked in the course of this thesis and the fact that the isomorphism check is the bottleneck in the algorithms is of course left unchanged. Figure 4 and Table 5 give an idea of the running time of Algorithm 2.[7]

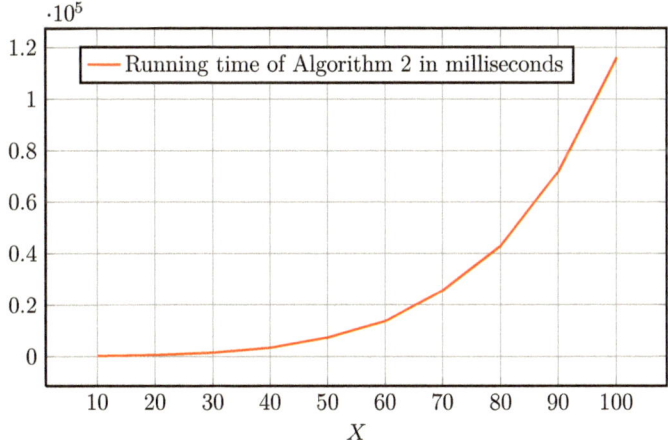

Figure 4.: Running time of Algorithm 2 in milliseconds

X	Time (ms)
20	421
30	1,357
40	3,229
50	7,099
60	13,510
70	25,490
80	42,900
90	71,652
100	115,815

Table 5.: Running time of Algorithm 2

---

[7] Running times for larger values of X have not been recorded. The running time for $X = 1000$ is however known to be several weeks and, as Figure 4 suggests, the running time of the algorithm is far from linear, quadratic or some other "fast" order.

6. Summary and Outlook

We have seen that the number of non isomorphic fields K with second successive minima $M_2(K) \leq X$ converges to $\kappa X^{5/2}$ where $\kappa = \frac{\sqrt{6}}{30}\zeta(5)^{-1} \approx 0.0787418966$. We have seen that the number of polynomials given by few constraints is asymptotically equal to the number of pairs (B, C) which can be considered as parametrizations of these polynomials. We have investigated how we have to modify, to "cut back" this set of pairs so that each number field that can be "generated" by an element of this set has—at least asymptotically—exactly one representative within the set.

We have computationally investigated the convergence of the number of fields with bounded successive minima of the trace form and have given an idea of the rate of this convergence. The convergence is actually quite slow as we can learn from the computational results presented in the appendix in Table 7 and Figures 5 and 6. For the first correct decimal place in the value of the fraction

$$\lim_{X \to \infty} \frac{|P_\kappa(X)|}{|F(X)|}$$

we have to choose a value of about 750 for X which already implies an original set of polynomials $P(X)$ of size $2,516,676$. The largest value for X which we were able to apply Algorithm 2 to is 1000.

To get an idea of the further development of the situation, Table 6 gives the sizes of the original sets of polynomials for larger values of X.

6. Summary and Outlook

X	$|P(X)|$
1500	14,233,175
2000	29,216,287
3000	80,506,393
5000	288,691,911
10000	1,633,039,795

Table 6.: Size of the set $P(X)$ for larger values of X

The following two figures give a vivid representation of the situation. It is obvious that $\kappa X^{5/2}/|F(X)|$ tends to 1 very slowly, as stated before while the behavior of the two graphs in Figure 5 provide evidence that the fraction actually converges to 1. The graph that figures the number of fields K with $M_2(K) \leq X$ consists in fact of points rather than being a continuous line. To provide better readability, however, the points have been linked by a smooth line.

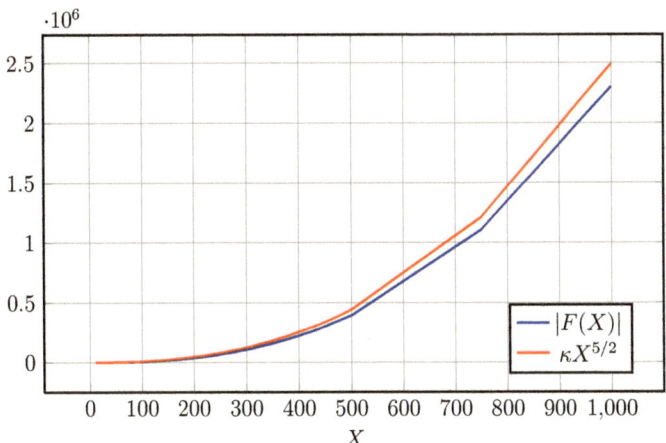

Figure 5.: Graphs of $|F(X)|$ and $\kappa X^{5/2}$

6. Summary and Outlook

What holds for Figure 5, also holds for the values of the fraction in Figure 6. The fractions have been computed only for some values of X but by linking them by a smooth line we provide much better readability.

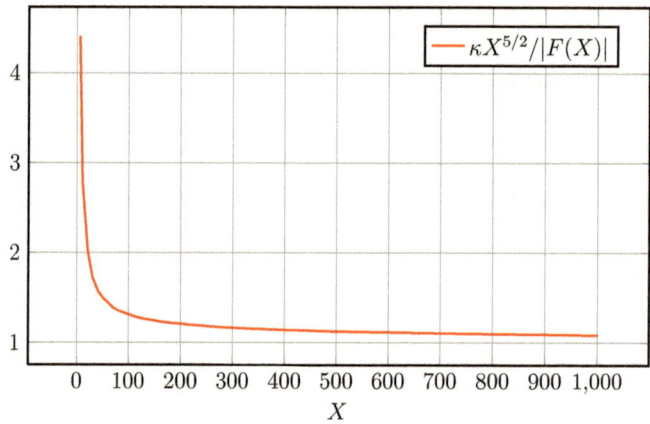

Figure 6.: Value of the fraction $\frac{\kappa X^{5/2}}{|F(X)|}$

We have seen that the reason for the relatively small maximum value of X which we have been able to investigate computationally is due to the bottleneck in the isomorphism-test for two given number fields.

An interesting extension of this thesis would be to investigate if there is a better isomorphism-test for number fields, either in general or for the special case of (totally real) cubic fields. Another possibility to take the discussion further would be to apply the theory to number fields of higher degree although it is not likely that a reasonable "universal" version of the theory can be built that suits all degrees of number fields.

Every new case introduces new individual questions and gives rise to new problems that have to be solved. For instance, we have used some results that work nicely with the case $n = 3$ but may not help that much (or maybe even will not work) for $n \geq 4$. Problems that have been solved by applying such

6. Summary and Outlook

results in our situation may not be that easy to solve in other cases or may require utilizing completely different methods and theories.

The theory of cubic fields and number fields in general with all their open problems as well as every subset of these problems will therefore stay one of the most vital fields in algebraic number theory.

A. Computational Results

The following table shows all information obtained by executing Algorithm 2 for computing the number of fields (subject to the choice of X) and ρ_X. The column "Ratio" contains the values of the fractions $\kappa X^{5/2}/\#\text{Fields}$ and the value of $\kappa X^{5/2}$ is rounded to the nearest integer.

X	#Pol.	#irr. Pol.	#Fields	$\kappa X^{5/2}$	Ratio	ρ_X	$\max d(K)$
10	62	22	9	25	2.778	4.737	473
20	309	192	70	141	2.014	59.086	3973
30	830	606	226	388	1.717	220.629	13473
40	1685	1330	507	797	1.572	549.433	31973
50	2930	2430	934	1392	1.490	1069.057	62473
60	4605	3936	1527	2196	1.438	1826.115	107973
70	6754	5908	2336	3228	1.382	2884.922	171473
80	9417	8376	3336	4507	1.351	4239.868	255973
90	12626	11382	4553	6051	1.329	5899.391	364392
100	16415	14948	6027	7874	1.306	7941.241	499973
110	20818	19116	7765	9993	1.287	10381.000	665473
120	25863	23918	9763	12421	1.272	13224.000	863973
130	31580	29378	12064	15173	1.258	16503.000	1098473
140	37997	35528	14611	18261	1.250	20168.913	1371973
150	45136	42394	17506	21699	1.240	24361.036	1687473
160	53027	49992	20744	25498	1.229	29146.526	2047973
170	61692	58364	24306	29671	1.221	34341.840	2456473
180	71155	67516	28142	34228	1.216	39990.105	2915325
190	81440	77490	32460	39182	1.207	46428.435	3429473
200	92569	88290	37038	44543	1.203	53288.154	3999973
210	104562	99956	42049	50322	1.197	60737.788	4630473

A. Computational Results

220	117447	112496	47453	56528	1.191	68852.614	5323973
230	131238	125944	53246	63172	1.186	77586.878	6083473
240	145959	140298	59464	70264	1.182	87009.534	6911973
250	161632	155614	66091	77814	1.177	97061.572	7812473
260	178267	171872	73095	85830	1.174	107619.799	8787973
270	195894	189124	80602	94322	1.170	119142.578	9841257
280	214525	207362	88546	103300	1.167	131242.700	10975973
290	234184	226624	96983	112772	1.163	144199.066	12194473
300	254883	246924	105772	122746	1.160	157612.104	13499757
310	276640	268266	115173	133232	1.157	172094.352	14895473
320	299477	290688	125016	144238	1.154	187369.939	16383973
330	323410	314198	135321	155772	1.151	203121.336	17968473
340	348457	338814	146106	167843	1.149	219775.224	19651892
350	374634	364560	157505	180458	1.146	237507.556	21437473
360	401957	391432	169270	193625	1.144	255724.196	23327892
370	430440	419472	181654	207353	1.141	274946.435	25326473
380	460103	448676	194476	221649	1.140	294875.243	27435973
390	490960	479066	207933	236520	1.137	315772.734	29659473
400	523025	510668	221952	251974	1.135	337806.188	31999973
410	556316	543484	236543	268019	1.133	360535.570	34460473
420	590847	577532	251358	284662	1.132	383398.315	37043973
430	626634	612838	267237	301910	1.130	408557.127	39753473
440	663683	649390	283338	319770	1.129	433656.673	42591973
450	702026	687232	300213	338249	1.127	460101.433	45562392
460	741665	726374	317485	357355	1.126	487149.504	48667973
470	782616	766814	335544	377095	1.124	515606.265	51911473
480	824897	808568	354096	397474	1.123	544890.629	55295973
490	868522	851682	373418	418501	1.121	575305.303	58824473
500	913503	896130	393105	440181	1.120	606152.540	62499973
750	2516676	2484474	1106971	1212995	1.096	1741855.204	210937473
1000	5165625	5115772	2300779	2490037	1.082	3661005.359	499999973

Table 7.: Computational results

B. Source Codes

We present the source codes of the PARI-implementations of the utilized algorithms as well as some further auxiliary programs. Some simple improvements of the codes have been made in some places although generally the implementations are straight forward and easy to follow if their sense is known. All source codes have been tested and verified with GP/PARI CALCULATOR Version 2.3.4[8] and may not work correctly with other, especially older, versions of GP/PARI CALCULATOR.

```
1   \\ compute the i-th successive minimum of the field defined by f
2   succmin(i,f) = {
3     K = nfinit(f);
4     return(K[5][4][i,i]);
5   }
```
Listing B.1: Successive minima

```
1    \\ compute the number of polynomials within the bounds defined by X
2    numofpols(X) = {
3      count = 0;
4      for(a = 0, 2,
5        for(b = ceil((a^2-X)/2), floor(a^2/3),
6          for(c = ceil(-1/27*(2*(a^2-3*b)^(3/2)+2*a^3-9*a*b)), floor(1/27*(2*(a^2-3*b)^(3/2)-(2*a^3-9*a*b)))),
7            count++;
8          );
9        );
10     );
11     return(count);
12   }
```
Listing B.2: Size of the set of polynomials defined by X

```
1    \\ compute the number of irreducible polynomials within the bounds defined by X
2    numofirredpols(X) = {
3      count = 0;
4      for(a = 0, 2,
5        for(b = ceil((a^2-X)/2), floor(a^2/3),
```

---

[8] GP/PARI CALCULATOR can be downloaded with the PARI/GP package as source distribution and self-installing binary at http://pari.math.u-bordeaux.fr/download.html.

B. Source Codes

```
6      for(c = ceil(-1/27*(2*(a^2-3*b)^(3/2)+2*a^3-9*a*b)), floor(1/27*(2*(a^2-3*b)^(3/2)-2*a^3+9*a*b)),
7        f = x^3+a*x^2+b*x+c;
8        if (polisirreducible(f), count++);
9      );
10     );
11   );
12   return(count);
13 }
```

Listing B.3: Irreducible polynomials

```
1  \\ compute the largest field discriminant among the polynomials in question
2  largestdisc(X) = {
3    ldisc = 0;
4    for(a = 0, 2,
5      for(b = ceil((a^2-X)/2), floor(a^2/3),
6        for(c = ceil(-1/27*(2*(a^2-3*b)^(3/2)+2*a^3-9*a*b)), floor(1/27*(2*(a^2-3*b)^(3/2)-2*a^3+9*a*b)),
7          f = x^3+a*x^2+b*x+c;
8          if (polisirreducible(f),
9            fd = nfdisc(f);
10           if(fd>ldisc,ldisc = fd;);
11         );
12       );
13     );
14   );
15   return(ldisc);
16 }
```

Listing B.4: Field discriminant

```
1  \\ compute the number of fields that are generated by elements of the set of polynomials defined by X
2  numoffields(X) = {
3    l = listcreate(10000000);
4    for(a = 0, 2,
5      for(b = ceil((a^2-X)/2), floor(a^2/3),
6        if(a==0,
7          cc = 0;,
8          cc = ceil(-1/27*(2*(a^2-3*b)^(3/2)+2*a^3-9*a*b)));
9        );
10       for(c = cc, floor(1/27*(2*(a^2-3*b)^(3/2)-2*a^3+9*a*b)),
11         f = x^3+a*x^2+b*x+c;
12         added = 0;
13         if(polisirreducible(f),
14           fd = nfdisc(f);
15           for(i = 1, length(l),
16             if(fd == l[i][2] && added == 0,
17               if (nfisisom(f,l[i][1])<>0,
18                 added = 1;
19               );
20             );
21           );
22           if (added == 0,
23             listput(l,[f,fd]);
24             added = 1;
25           );
26         );
27       );
28     );
29   );
30   return(length(l));
31 }
```

Listing B.5: Number of fields

B. Source Codes

```
1   \\ compute the number of fields that are generated by elements of the set of polynomials defined by X and
      compute the corresponding value of rho_X
2   numoffieldsrho(X) = {
3     l = listcreate(10000000);
4     rho = 0;
5     for(a = 0, 2,
6       for(b = ceil((a^2-X)/2), floor(a^2/3),
7         if(a==0,
8           cc = 0;,
9           cc = ceil(-1/27*(2*(a^2-3*b)^(3/2)+2*a^3-9*a*b));
10        );
11        for(c = cc, floor(1/27*(2*(a^2-3*b)^(3/2)-2*a^3+9*a*b)),
12          f = x^3+a*x^2+b*x+c;
13          added = 0;
14          if(polisirreducible(f),
15            K = bnfinit(f,flag=2);
16            fd = K.disc;
17            for(i = 1, length(l),
18              if(fd == l[i][2] && added == 0,
19                if (nfisisom(f,l[i][1])<>0,
20                  added = 1;
21                );
22              );
23            );
24            if (added == 0,
25              hF = K.clgp.no;
26              RF = K.reg;
27              rhoF = 2^(3-1)*hF*RF/(fd^(1/2));
28              listput(l,[f,fd,hF,RF,rhoF]);
29              rho += rhoF;
30              added = 1;
31            );
32          );
33        );
34      );
35    );
36    return([length(l),rho]);
37  }
```

Listing B.6: Number of fields and ρ_X

References

Apostol, T. (1998). *Introduction to Analytic Number Theory*. Springer, Berlin, corr. 5th edition.

Belabas, K. (1999). On the mean 3-rank of quadratic fields. *Compositio Mathematica*, 118:1–9.

Bhattacharya, P. B., Jain, S. K., and Nagpaul, S. R. (1994). *Basic Abstract Algebra*. Cambridge University Press, 2nd edition.

Buchmann, J. and Ford, D. (1989). On the computation of totally real quartic fields of small discriminant. *Mathematics of Computation*, 52(185):161–174.

Buchmann, J., Ford, D., and Pohst, M. (1993). Enumeration of quartic fields of small discriminant. *Mathematics of Computation*, 61(204):873–879.

Cohen, H. (1993). *A Course in Computational Algebraic Number Theory*. Springer GTM, New York.

Cohen, H. (2000). *Advanced Topics in Computational Number Theory*. Springer GTM, New York.

Cohn, H. (1954). The density of abelian cubic fields. In *Proceedings of the American Mathematical Society*, volume 5, pages 476–477. AMS.

Davenport, H. and Heilbronn, H. (1969). On the density of discriminants of cubic fields. *Bulletin of the London Mathematical Society*, 1:435–348.

Davenport, H. and Heilbronn, H. (1971). On the density of discriminants of cubic fields ii. *Proceedings of the Royal Society of London*, 322(1551):405–420.

References

Ellenberg, J. S. and Venkatesh, A. (2007). Reflection principles and bounds for class group torsion. *International Mathematics Research Notices*, 1.

Freundt, S. (2005). QaoS - Querying algebraic objects System. Web interface to the QaoS database. Available at http://qaos.math.tu-berlin.de.

Hasse, H. (1930). Arithmetische Theorie der kubischen Zahlkörper auf klassenkörpertheoretischer Grundlage. *Mathematische Zeitschrift*, 31(1):565–582.

Helfgott, H. A. and Venkatesh, A. (2006). Integral points on elliptic curves and 3-torsion in class groups. *Journal of the American Mathematical Society*, 19:527–550.

Hungerford, T. W. (1974). *Algebra*. Springer GTM, New York.

Lenstra, A. (1983). Factoring polynomials over algebraic number fields. In van Hulzen, J., editor, *Computer Algebra*, volume 162 of *Lecture Notes in Computer Science*, pages 245–254. Springer.

Marcus, D. A. (1977). *Number Fields*. Springer, New York.

Milne, J. S. (2009). Algebraic number theory (v3.02). Scriptum. Available at http://www.jmilne.org/math.

Pierce, L. B. (2005). The 3-part of class numbers of quadratic fields. *Journal of the London Mathematical Society*, 71(2):579–598.

Pohst, M. (1975). Berechnung kleiner diskriminanten total reeller algebraischer zahlkörper. *Journal für die reine und angewandte Mathematik*, 278/279:278–300.

Pohst, M. (1982). On the computation of number fields of small discriminants including the minimum discriminants of sixth degree fields. *Journal of Number Theory*, 14:99–117.

Pohst, M., Martinet, J., and Diaz y Diaz, F. (1990). The minimum discriminant of totally real octic fields. *Journal of Number Theory*, 36:145–159.

References

Pohst, M. and Zassenhaus, H. (1989). *Algorithmic Algebraic Number Theory*. Cambridge University Press.

Roberts, D. P. (2000). Density of cubic field discriminants. *Mathematics of Computation*, 70(236):1699–1705.

Salmon, G. (1960). *A Treatise on Conic Sections (AMS Chelsea Publishing)*. Chelsea Publishing Company.

Shintani, T. (1975). On zeta-functions associated with the vector space of quadratic forms. *Journal of the Faculty of Science, The University of Tokyo*, 22:25–66.

Wong, S. (1999). On the rank of ideal class groups. In *Proceedings of the Fourth Canadian Number Theory Conference*, pages 377–383. AMS.